P9-CNH-122

MIRROR TOUCH

MIRROR TOUCH

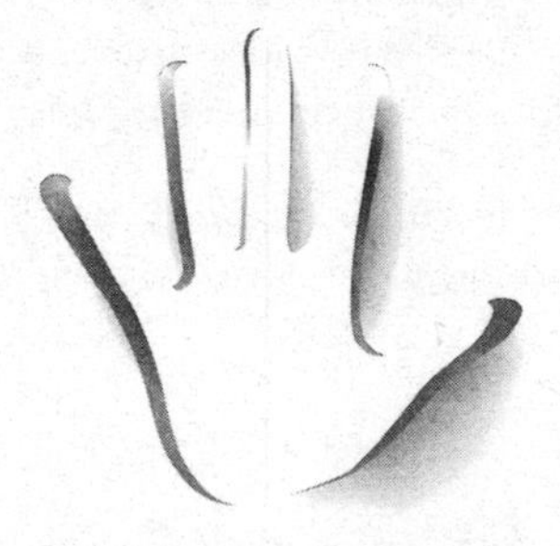

NOTES FROM A DOCTOR WHO CAN FEEL YOUR PAIN

JOEL SALINAS, M.D.

HarperOne
An Imprint of HarperCollins*Publishers*

HarperCollins books may be purchased for educational, business, or sales promotional use. For information, please email the Special Markets Department at SPsales@harpercollins.com.

FIRST EDITION

Designed by Beth Shagene

Library of Congress Cataloging-in-Publication Data has been applied for.

ISBN 978-0-06-245866-7

17 18 19 20 21 LSC 10 9 8 7 6 5 4 3 2 1

To my parents, to my brother and sister, and to those who entered my life as patients, teachers, or both.

Table of Contents

INTRODUCTION

Sensorium

I BETRAY MYSELF.

During my first week as an internal medicine clerk, I was going through my patient list with the attending on duty when a code blue was called in the waiting room. Before the announcement on the overhead system finished ringing out, the attending and I were already bolting out the door. This was my first emergency, and I was eager to run through it.

Right around the corner from our office, a man was unconscious on the ground. His wife cowered in the corner, horrified. A young male nurse in dark blue scrubs pulled in a crash cart, which rattled with stocked emergency supplies. One of the medical residents started compressions. All around me were shouts of "P.E.A.! P.E.A.!" I could only make out a few of the medical directives in the alphabet soup yelled over the chaos. I took note of each abbreviation as best I could to keep up.

Or, at least I tried. I was absorbed in the man in cardiac arrest, fully immersed in his bodily experience. The sensations in my body mirrored the sensations in his. Compression after compression on his chest and on mine. I felt my own vocal cords tighten as doctors slid a tube down his throat—a sharp object shoved down the back of *my* throat. I told myself over and over that the man would be alright, that we would save him. Because that's what we do in medicine. We save people. After we save him, I assured myself, we'll discuss what worked, steps and procedures I can repeat the next time someone needed to be saved. But, as the doctors continued chest compressions, I felt my back pressed firmly against the linoleum floor, my limp body buckling under each compression, my chest swelling with each artificial breath squeezed into me through a tube, a hollow slipping sensation.

I was dying, but I was not.

After thirty minutes, the hospitalist called the code. The man's wife let out a wail, which was soaked in frayed black and oak. I stared intently at the dead man's body. I couldn't move. I lay there with him, dead. The absence of sensations in my own body, the absence of movement, the absence of breath, a pulse, any and all feeling. In my body, nothing but a deafening absence. I had to step away. I had to will myself to breathe.

I escaped into the nearest bathroom where I fell to my knees and felt the rest of my body rush toward my face. I vomited. I vomited until I started dry heaving. I was alive, though it felt as if I had died. I felt it, without a doubt, as sure as I could feel the tears and saliva currently spilling out of me in the bathroom stall. The contradiction brought my stomach into a knot once again. I had to stop. The rest of the team was going to wonder what had happened to me.

I took in a long deep breath. I flushed the toilet along with all of the turmoil that I had just expelled and stared at my reflection in the water. As the water settled, so did I. Another deep breath. I stood up and washed my face. I stared at myself in the mirror.

This was me, I told my reflection, this was my body. I slowly felt embodied again, becoming in tune with the sensation of my clothes against my skin, the position of my arms and legs, the weight of my own flesh suspended on my bones, my heartbeat, my pulse, my rising and falling breaths. The hum of the paper towel dispenser was the last sound I heard before I finished wiping away what had happened. I was not going to let this happen again. Patients were going to rely on me. And I needed to be there for them. I was going to endure this and make it past pain, past death, past all the suffering I experienced to reach down and treat them in the throes of their illness. To process the echo of another's darkest and most desperate anguish was my responsibility after I went home. I took one last deep breath as the door closed behind me.

My name is Joel Salinas. I am a neurologist and a polysynesthete, a person with multiple forms of *synesthesia*. Through *mirror-touch* synesthesia, one form of synesthesia, my body physically feels the experience of others—sometimes betraying me, losing me in the people I see before me.

At the conscious level, I feel the physical sensation of touch while observing it occur on another person. I am aware of the mental process taking place and can even describe the sensations in detail. Triggered automatically by sight, I feel a mirrored touch on parts of my body that correspond visually to whom I'm looking at face-to-face—your left, my right, your right, my left—like in a mirror. But when I am standing next to a person, side-by-side, the location of my synesthetic touch is more likely to be anatomical—left-to-left, right-to-right—as if we were in the same body.

Walking through the hospital's revolving door, for example, I see an elderly woman in a wheelchair. She's dressed in a large worn tweed coat with dark floral embellishments and a burgundy knit hat covering a nest of gray hair that slips out from under the rim of the hat: I feel the sensation of vinyl pressed against the back of my thighs, sitting bent at the waist, sinking into the seat and coat, the snug fit of the woman's hat around my forehead and scalp, my fingers laced together in front of my chest. I feel the movement of her eyes and eyebrows as she looks through the glass doors, then rests her gaze back down at the ground. Holding on to the handles behind her stands a volunteer. He's next to the metal pole that extends from the wheelchair with a lime green plastic bag of belongings suspended like a bindle. Wearing an oversized salmon-colored shirt, the uniform of volunteers, the volunteer leans with his hip jutted toward his right: I feel the phantom contraction of muscles in my left hip, his glasses sitting invisibly on the bridge of my nose. The acne on his face peppers my own cheeks. Passing by the security guard at the welcome desk, I raise my ID badge: I feel the coiled plastic earpiece tucked around the security guard's right ear around my left, the weight of his black wool suit on my arms and shoulders, the height and width of his towering frame, the fatigue in his eyes. Entering the long express line for early morning coffee at the hospital's main café, I see other doctors, nurses, therapists, patients, custodians, administrators: each is a different channel, a different collection of feelings and expressions. My brain's perception of bodily sensations surfs across them as my eyes pass over each. I line up behind a mother holding her baby over her left shoulder. She rocks from side to side. I feel the weight of the baby against my left shoulder, the tender sway of my torso rocking from side to side, the bob of her hair brushing up against the back of her neck. I feel the baby, the roundness of our faces, the clutching of our tiny hands. We look at each other. The baby smiles and I feel the smile on my face, and then my own smile realizes it.

While mirror-touch synesthesia is relatively rare, the most

common form of synesthesia is likely *grapheme-color* synesthesia where each grapheme (the collective term for all written forms of numbers and letters) evokes the experience of a specific color. Regardless of the actual color of what's written, I simultaneously see its synesthetic colors layered cryptically on top. In the individual letters of the word *cat,* for example, C is black, A is red, and T is red-orange. And collectively, the word cat evokes clouds of color from each letter, which appear as a nebulous black haze with puffs of red and Montana dust.

My synesthesia also goes beyond neurologically colored numbers and letters. I can perceive motion as sound, music as color, taste as shapes as well as a wide variety of other exotic manifestations. About 4 percent of the general population have some form of synesthesia, including the physicist Richard Feynman, and the music producer Skrillex. Though synesthetic traits can exist in anyone, there seems to be a historically higher prevalence of synesthesia among artists. The ranks of legendary musicians with synesthesia include Jimi Hendrix, Stevie Wonder, Billy Joel, Tori Amos, and Eddie Van Halen, to name a few. Through its vast combinations of sensory associations, synesthesia erodes the barrier between the mundane and the electric, the predictable and the unknown, granting musicians, writers, artists, and cultural innovators the ability to share their synesthetic world in provocative ways. Franz Liszt, for example, famously conducted his orchestra with requests for more violet tones, while Marilyn Monroe reportedly saw colorful vibrations with sound. In his autobiography Vladimir Nabokov explained in vivid detail his correlation between colors and the alphabet, gorgeously describing the letter S as a "curious mixture of azure and mother-of-pearl." I, on the other hand, experience the letter S like a fall squash, an autumnal mixture of yellow and orange, an amber-tone sibilant.

A more complicated layer of my synesthesia is referred to as *ordinal linguistic personification*. In my case, every grapheme has not only a unique color but also unique human characteristics. This is especially true of numbers. I consider numbers familiar friends. They exist in my world with innate and diverse personalities, like the number 3, a modest indigo that shies from its true potential. And in reverse, everyone I meet is tied to a synesthetic experience that is similar to what some might call an aura, where everyone is connected with images of at least one color, which is instantly linked to his or her corresponding numbers in varying sizes and configurations, creating a mosaic of colored numbers. A friend from medical school, for example, is a great big turquoise 7, eclectic yet endearing, surrounded by a smattering of chartreuse 6s, each weirdly awkward, and a halo of cool blue 4s, friendly and pacific. As I got to know him, his graphemes multiplied, diversified, and populated into a grander pictorial arrangement. The gradual accumulation of personal information—or data points, in scientific parlance—settled into an image of a large translucent lake nestled in a pale gray crater with turquoise shores and a light teal-colored center (Pantone 3245, to be precise).

Layer upon layer, my synesthesia blends. People with azulean 4s are alluring. A bite into a strawberry, barely ripened, turns my world into a splash of aqua filled with the sound of crashing cymbals. The trilling clarinet of *Rhapsody in Blue* never fails to evoke a bright, slithering, serpentine figure at the base of my tongue that tastes like blueberries perfumed lightly with fresh tire tread.

This is not a drug-induced daydream. This is my reality.

It occurs at all times with all my senses; my mirror-touch sensations are active even when I'm not facing a living person. Standing before the statue of *David,* for instance, I feel the tension of my left sternocleidomastoid muscle as if I were turning my head toward the right. I feel, too, the heavy sensation of thick cloth on my right shoulder, the lightness of my left hand, and the slight sensation of my right knee bending inward. When looking at the Statue of Liberty, I feel the heavy cloth draped across the dorsa of my feet, the sensation of holding weight in the crux of my right arm and hand, the tension of my left triceps and shoulder as if I were, like Lady Liberty, extending my arm upward. The pointed array of horns emerges from my scalp just before the midcoronal plane. I am their living mirror.

I experience phantom physical sensations at more rudimentary levels of visual information that do not contain potential for human facial expressions. Looking at a glass of water, for instance, I automatically feel a tickle up along the corners of my mouth as if peeking my head just above the water's surface from the mouth up. A handbell elongates my upper body while my lower body becomes frictionless and spacious. Lampposts stretch me upward with my head positioned high up top. Electrical outlets, with their mousey look of surprise reflected on my own face, make me feel friendly and mischievous. I can recognize this experience as a combination of pareidolia, the phenomenon of recognizing familiar patterns such as faces where none exist, and apophenia, our brain's instinctive reflex to make sense of random information.

I feel features of basic visual characteristics, too, such as sharp angles, rounded edges, and contrasting colors, all playing into my spontaneous vivid emotional experience, empathizing with whatever emotion I subconsciously project onto the world around me. As such, any observed piece of art can be imprinted onto my body, sculpting me into an extension of the artist's creation. Appreciating a navy and white Chihuly glass sculpture, for example, I feel not only spikes pushing against my body but also spikes pushing outwardly up through my skin as I slowly, surely, become the piece. It is as if I were transforming into an arctic sea urchin—cold, erratically defensive, drenched in fearful distrust. Even the grid of a square lattice leaves a tactile imprint that feels as though

I were literally leaning my face against a screen door, conjuring in me a feeling of tension and the quiet frustration that comes with longing to escape from a suffocating prison.

Such experiences are primarily a sensory process, but their volume, their *realness,* can also be affected by higher-order cognitive functions. An acute attention to detail, coupled with a heightened sense of awareness and personal significance, seems to drive the salience of my synesthetic experiences. The sight of spine-covered bark is particularly vivid, probably due to an incident playing tag as a child. Growing up in South Florida, a game of tag outdoors usually meant that the trunk of a nearby palm tree was "base." I still remember when I mindlessly slapped my hand on the bark of one of these palms to escape being tagged. I shrieked and yanked my hand backward to reveal the palm's quill-like spines lanced into the skin of my palm. Because of the surprise and heightened emotional charge from the pain of this memory, when I see spiny palm trees today I experience the intense sensation of invisible spines over my face as if I were rubbing my face against its bark.

Extremely vivid synesthetic experiences always have the potential to be exquisitely challenging. Rare or unexpected situations make it almost impossible to differentiate between objective physical reality and my own internal subjective reality. In the hospital, while I examine a condition or perform a procedure for the first time, like threading a long wire into the chest, the likelihood of experiencing the sensation or pain as if it were actually occurring to me is increased significantly. During my neurology training, as I began to see patients with Tourette's syndrome and tic disorders, I distinctly recall one patient who in the setting of significant stress developed new self-mutilating tics. He would chew on the insides of his mouth and push the corners of his lips with his knuckles so hard that his cheeks split apart like shredded beef. Watching him chew on the flesh of the right side of his face while grinding his teeth with all his force, I felt a painful buzzing run through the left side of my face and mouth that was so vivid that it bordered on hallucination. It was as if a stun gun were pressed against my face and triggered with each of his tics. The more forcefully he pushed, the more vivid the pain. The mirror-touch sensations are constant. But in these instances, they pierce through my ability to filter, invading my perception of reality.

Calling synesthesia a neurologic *disease, disorder,* or *condition* is a bit of a technical misnomer because synesthesia, as a whole, is not considered an independent source of significant social or functional impairment. In the absence of well-defined pathology, I prefer to call it a *variant* or *trait,* a neurologic feature with the capacity for good and bad, strengths and weaknesses, depending on the circumstance—much like some of us easily soak up new languages but collapse under the arithmetic weightlifting of calculating tax on a restaurant bill.

Existing in a neurologic replica of another's sensory perceptions is as close as I can get to literally putting myself "in the other person's shoes." It is then up to me to walk the mile in front of me. In other words, if empathy is a person's capacity to understand and feel the experiences of another person from their perspective, then mirror touch represents a persistent heightened state of empathy, the potential for a more fully realized form of empathy. The limit of empathy is, of course, that we're not the same as another person. We do not share the same body. Nor do we share the same mind. Thus, we might assume that, because we're not the same person as the other, his or her perspective perhaps merits less value—that it is less worthy of our attention or awareness. It becomes easy, almost natural, then, to assume that another person's experience is already too different from our own to empathize with or understand in full. Our unconscious reflex might be to make no more than a passing attempt to experience and understand another person's perspective. After all, we have our own issues, concerns, and pains, so why expose ourselves to more discomfort, more of what we already want to avoid?

With mirror touch, my decision to work toward empathy is automatic, compulsory. Though it doesn't reveal itself in full, it does offer the *potential* for a more fully realized empathy. I still need to interrogate these mirrored feelings and sensations, asking questions and engaging with these feelings and sensations to determine their significance. Conducting such internal interrogations, as Isabel Wilkerson writes, calls for a "radical empathy," which occurs when we put ourselves inside the experiences of others and, in due process, allow an honest and authentic re-creation of their joy, their pain, their suffering, whatever they are experiencing, within ourselves.

My trait, though, can just as well blur the boundaries between myself and those around me to the point that I can become inextricably entangled in others—their emotions, their needs—at the expense of losing myself in other people. For as far back as I can remember, information has constantly poured through my mind. Creating a mental filter for the sake of self-preservation may sound simple, but it is a dangerous means of survival. Filter too much and I risk numbing my senses completely and, as a consequence, forfeiting my humanity, my ability to feel and to empathize. Filter too little and I risk submerging myself too deeply in another person, drowning in my own senses and losing all sanity, all sense of myself.

This filter was brought to life when I was sitting at the head of a hospital conference table leading a family meeting as the senior resident responsible for the inpatient neurology service. I was surrounded by nurses, therapists, a social worker, a case manager, and a dying patient's family. The patient was an older woman with severe dementia who had fallen ill with a barrage of infections, strokes, and seizures. Her family was not ready to let her go. Her eldest daughter was the most adamant in this regard. The family wanted to hold on to her as long as medical technology could allow while many members of the woman's medical team felt only guilt and distress about causing her more suffering during her inevitable descent into death. Emotions were high. I had to make continuous adjustments to keep the focus on moving the conversation forward, guiding the conversation back on track, while honoring the various emotions around the table. Expressions and gestures of grief and anger filled the room. The facial expressions of the family pulled me into their torment and confusion: I felt my brows furrow, my eyes widen and dart around the room looking to anyone who could give an answer, a simple, satisfying directive that did not exist. I became them, the entire family, just as I'd lost myself with the rest of the room. The head of the table may as well have been an empty chair.

Pulling back, however, bringing myself back into my feet, into my skin, I could once again follow the expressions and gestures closely enough to know when each person was ready to erupt in emotion or needed to speak, vent, or simply feel acknowledged. I felt when a person was ready to contribute. Buttressed by what medical knowledge I had to offer, I reflected back what had been reflected in me, what the group felt. My body felt a brief moment of serenity as my eye caught a family member who had been sitting back quietly and reflecting, the younger sister. She was ready to share. I spoke her name, an open invitation. She looked toward her sister and said with a calm momentum, as if on cue, "I love mom, too. And I know this hurts so much for both of us, but I think we both know what mom would've wanted." The older sister replied by placing her hand on top of her younger sister's, a quiet acknowledgment of compassion and gratitude. I felt the older sister's shoulders lower, her breaths become longer. Together, they decided to honor their mother's wishes. Together, as a group, we decided to let the woman go.

I always *know* my body's physical form, where I end and someone else begins. I can feel the actual touch of my clothes, the pressure of the floor against my feet, the nerves in my joints telling me how and where my body is in space. However, another layer of internal sensory experience co-occurs within me. It sends contradictory information through my brain, top-down, in direct competition with the almost undeniable bottom-up information that informs me where I am and how I stand. Add to these experiences all my other synesthetic associations, compounded exponentially in interlaced layers of sensory perception, and my daily existence may start to seem as if I'm seeing the world through a kaleidoscope, gazing into an endless multisensory landscape, while living a deep, indefinite dream—distant and at odds with rational scientific thought.

Each person employs his or her own collection of remembered and experienced perceptions, his or her own set of lenses through which to view his or her external and internal worlds. Perhaps this is what makes empathy so challenging—and so compelling. At its core, empathy seems to require an initial spark of *desire* to switch perspectives, to generously give another person's experience enough worth so that we're not only willing but also *yearn* to see and live the world through that person's perspective. Mirror touch may provide some clues into what we need to make our ability to do so a reality. If we could better understand and harness the experience of synesthesia, particularly mirror-touch synesthesia, what could it possibly teach us about the brain, about ourselves, and about our ability to connect and remain, as Eula Biss describes it, "continuous with everything here on earth, including, and especially, each other"?

The emerging field of synesthesia research has only just begun to probe the brain, which largely remains a black box. The first scientific descriptions of synesthesia occurred in the early nineteenth century, initially as case studies. Because of the inherent challenges in measuring subjective experiences at the time, compounded by the rise of behaviorism in psychology, the study of self-reported subjective experiences fell out of favor in almost all scientific communities. It was not until the early 1980s that pioneering researchers turned once again to the scientific investigation of synesthesia, including neurologists like Richard Cytowic who were emboldened by the cognitive revolution and novel brain imaging tools. By the late 1990s, while I was still in high school, V. S. Ramachandran and other neuroscientists sparked a major resurgence in the study of synesthesia, providing sufficient objective evidence to support that synesthesia, once considered nothing more than a subjective curiosity, is a genuine, measurable sensory experience anchored in tangible neurobiological mechanisms.

Today, as new tools for studying the brain become available and more researchers realize the tremendous value of this phenomenon, the field of synesthesia research continues to grow. Perhaps someday even my murkier subjective experiences can be explained empirically, thoroughly categorized by their underlying biology, just as knowledge over the years has evolved for other neurologic traits.

Mirror touch has been a harsh, but just, teacher. Since childhood, my trait has required an almost monastic dedication on my part to the physical and mental labor of slowing down or filtering out the tides of sensory information while preserving my dauntless, bloodied pursuit of curiosity. Through my trait's many humbling and unexpected lessons, I've cultivated a greater awareness of our shared humanity, a deeper understanding of other people, and a truer sense of where we all begin and where we all end.

But not without sacrifice. And not without a struggle.

My hope is the following pages will serve as my case history of mirror touch and my various forms of synesthesia in order to share what I've come to realize throughout this living labor. This book, then, is a collection of my experiences from my childhood through the present—everything, in short, I've learned about the trait in my professional and personal life and, in the process, everything I've come to understand about myself and the human condition, and everything it means to think, to feel, and to be—as I have lived them, through myself and through others.

This is my story. These are my experiences.

CHAPTER 1

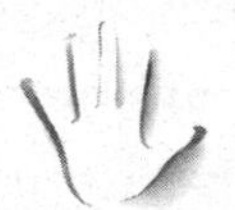

Where I End and You Begin

One of my earliest memories, long before I knew about synesthesia or mirror touch, involved coloring, and the confusion and concern it caused me. I remember sliding my tiny hands across a workbook whose thick, coarse pages were constructed to instill a basic understanding of shapes in me and the rest of my kindergarten classmates, five-year-old offspring of native Floridians or freshly immigrated families.

Though small, my hands could still appreciate the dimples lining the heavy paper. Loose perpendicular strands from the carelessly woven fibers tickled my fingers like kitten whiskers. Quietly, I brought my face down and brushed my nose and right cheek against the page. Its hard, black ink was embossed to corral savage fingers like mine into prescribed spaces, an early temptation against the tyranny of lines.

With my finger I traced the shadow of invisible images, experiencing them as living things, though they remained mysterious and elusive.

"See?" I remember telling my classmate who had the misfortune of sitting next to me. "There's a mountain here with the face of a wolf opening its mouth on it. And here. See the green baby dragon? It has pink hair and orange horns, and it's asleep."

My classmate stared at my finger delicately tracing the blank space. He squinted briefly, unsure of what he was supposed to look at, then quickly turned away to find something more interesting to do. I didn't pay his response much mind, though. After all, I had the shadows, which bounced against the nooks, crannies, and valleys of the page as distinct as figures. They were like birthmarks in motion. Colors covered every inch of the page, but they weren't laid out like the ink on the cover of the workbook. This much I could determine, even at a young age. Instead, the colors on the page were like memories of color, firework trails in the tight spaces between the solid lines. The colors flickered so quickly that they appeared to shine as a single color. If I shifted my attention to a connecting nook, the entire image changed, flickering in a new light and creating in the process a grander, more illuminated image. Dogs quickly transformed into dragons, which suddenly turned into damsels. If I wanted to make the pictures visible to others, I had to hurriedly trace the edges of each discovery with a crayon or pencil before the image changed again.

I could spend all day bringing images to life. But I wanted to be a good student, which meant following the teacher's instructions. "Just color in the lines, Joel," my teacher reminded me. But which lines? Which color? I wanted to color the lines within the lines. I wanted to color outside the lines, too, where there were infinite lines to color that would squirm, expand, and vanish, depending on the thoughts that came to my head as I studied their living contours.

"Done!" squealed another student with delight. Was I just slow? Or were my classmates blind to how much *could* be colored?

I saw chaotic green-yellow scribbles on my neighbor's page. A hundred nauseated curls violating a circle. It looked like tangled yarn gushing over the edge of an empty paint can. That's not right, I thought. The circle is obviously a darker red. A navy blue or a red-orange could work, I assured myself, but the circle was way too big and heavy for a green-yellow crayon. This was inexcusable.

I looked back at my page. There were too many colors and images to uncover. But then again, who would appreciate it? It mattered to me, but it didn't matter much to my teacher. I just had to find a way to get through this dilemma to earn a Gold Star, a telltale sign of being "good." I removed the tortured blue crayon from the Crayola box and moved it firmly along the inside of the black line. As I did so, the shadowy figures receded from the page, chased away by the presence of a prescribed shape. With the distractions fading into the background, I was free to sweep the crayon lightly back and forth until the entire circle was filled in with blue.

Maybe if I colored myself in this way, I thought, everyone would start to see the same things I saw, or barring that, I would start to learn how to focus my attention only on what everyone else saw.

I knew at an early age I was different, even though I didn't know why or how I was different. In school or at home, my behavior was never what you would call normal, or straightforward, or even easy.

Dressing me was a particularly arduous daily task for my mother. To avoid the scraping of clothes against my skin, I would constantly undress and stand naked in the middle of my bedroom until my mother forced me to put my clothes back on. It was unbearable and painful, and shirt tags were the worst. Getting me ready for school one morning, my mother pulled a new pink and gray striped shirt over my head. It was a gift from my grandmother. The material was heavier than the thin material I had learned to tolerate. The cloth around the collar was thicker; the seams around the sleeves were rougher on my bare skin. The tag was large, coarse, and folded over with a pinch. Even standing still, I felt it against my neck, as if a matchbook had been pressed into my collar. I began to whine and whimper. The edges of the tag against my skin felt like a hermit crab clawing at my neck. I flailed my arms backward and desperately tugged at the shirt to pull it out. But the more I struggled, the more the hermit crab dug its claws into my flesh. My mother quickly slipped the shirt off me, asking me what happened. "The thing," was all I could say, pointing at the collar, huffing between aggravated sobs. Later, my mother cut the tags off all my shirts.

I was different in other ways, too. Whenever I got excited, the feeling of exhilaration bubbled over the confines of my mind and into my body. I would flap my hands as if my very fingertips were lit Roman candles. My cousins called it "bird power" and eventually came to treat it like a good luck charm, which came in handy whenever they played Nintendo and reached the final boss's lair at the end of a video game. And then there was my love of colored Fisher-Price alphabet refrigerator magnets. While nearly every child my age loved to play with these magnets, I was obsessed with them, nearly fanatical. My favorite magnets by far were A, E, H, and U because their actual colors—unlike my unpleasant experiences with coloring books at school—matched their "right" color, the letter's inherent color only I could see. The "wrong" color, as far as I was concerned, was an insult to the dignity of the letter, a grave defamation of its character. When the Fisher-Price-painted colors matched the colors I saw for each letter, it was as if my external and internal landscapes became aligned, and I felt a pleasant and unexpected relief, like scratching an itch I didn't know I had.

To my teachers, I was the perfect student. Quiet. Diligent. Obedient. To my classmates, however, I was a nuisance. Serious. Withdrawn. Weird. While they were joyous, rambunctious, normal, I was a jigsaw piece from a completely different puzzle. Every kid my age always seemed uncomfortable around me, including my brother, Rainier, who considered me a regrettable nuisance inflicted upon him—his weird, fussy, and overly concerned older brother. My parents always told me to be myself. But for some reason, being myself only got in the way of making friends. After a few minutes with me, the expression on a classmate's face usually contorted with repulsion, confusion, or worse, fear. Perhaps the reason I didn't have many friends was the fact that I was too forthcoming about wanting to hug everyone, to be touched, to be *real* friends . . . I loved hugs. Not because I was especially needy or deprived of love at home. To me, the physical act of hugging was a wholly immersive experience; it overwhelmed me. Embracing another human being made me feel warm and secure, just like everyone else. But unlike everyone else, hugging also made me feel cool silvery blue, the same feeling the number 4 inspired in me. Whenever I hugged someone, I felt an immediate corporeal relief. Tension would slough from my body, and the world would start to make much more sense, just as it did when colors matched their letters. A physical touch from another human against my body assured me, physically, that we would take care of each other, that everything was right with the world. The experience was that visceral to me, that tangible. Looking back, I understand much of this relief came from how this simple physical expression of love or affection reduced the dissonance between the external world and my internal world, which was so alien from everyone else's. Because I constantly tried to hug them, my classmates teased me or ran away from me, shrieking.

My eventual solution was to dig a hole under the lowest wooden beam of the playset, a hole wide enough to slide into, like a dog trying to claw its way under a fence. The cracked wooden beams supporting the platform were stacked close enough that only a few streams of light poured in. My classmates couldn't see me. And if they couldn't see me, they couldn't avoid me or ridicule me. Underneath the cool shadow of the playset, I could hear the hollow thumping of my classmates' feet against the dried wood accompanied by their laughter and high-pitched prattling. With every one of their steps, the entire platform trembled, and sand trickled down through the cracks between the boards. I sat in silence, hugging my legs close to my body, pressing my cheeks and forehead into my knees.

I didn't always feel the sting of separateness. If I needed to, I could turn to my usual company. Not from people per se but from *things,* inanimate objects. I was always surrounded by a world heavily populated by everyday objects and miscellaneous things. Every aspect of the world around me was alive, animated with a distinct and fully formed personality. My eyes and my mind infused every *thing* with life, with real emotions—though fortunately these things and their emotions were less capricious than people. I appreciated these living objects for their individual histories, and I responded to each in much the same way as I would respond to a person. Almost anything I saw or touched around me was open and kind, available. We befriended one another, adored each other, even. My first-grade teacher once commended me for keeping my school box and school supplies in such good condition. But in truth, I only took care of my things because my pencils, pens, crayons, and even my Elmer's glue bottle each had its own life to preserve, its own background to acknowledge, its own right to coexist unharmed. At home, for instance, if I wanted to feel enveloped by a lively trickster, I would hop into my grandmother's bright maple rocking chair, which was young and spritely, especially in comparison to the loveseat, which (with its floral seat covers and mahogany bun feet) was somber and schoolmarmish. If I wanted to feel secure, more certain of the ground under my feet, I climbed onto the loveseat, falling into its earnest, stately contours.

I recall watching my grandfather pulling the morning edition of *El Nuevo Herald* out of its translucent yellow plastic bag. Before discarding the bag, he always tied it in a knot, sometimes three. My favorite was when he would tie two knots close together at the edge of the yellow plastic. Unbeknownst to him, he had created a living thing, like plastic *shikigami,* a girl with a ponytail and a long gown draping her below her neck. These bright yellow plastic knotted girls were mental concoctions generated from the combination of pareidolia and childhood imagination. But there was a certain elusive *something* that breathed life into them, the same elusive something that breathed life into other inanimate objects—like lightning sparking Frankenstein's monster to life. To me, these objects all conveyed real emotions, real personality. As hard as it is to believe, I made tangible human connections with them; they were as real as any other relationship in my life. I incorporated these plastic *shikigami* into my toy collection, along with other objects that had been tossed away. Each toy, every scrap, imbued me with thrills of emotion while I imbued it with a pulse. Like most children my age, I played with these objects by building them into elaborate scenes. I was the producer, director, stunt coordinator, dialect coach, and master puppeteer. Trash and action figures were my venerable cast. And, like many children, the themes were reminiscent of those that I had grown accustomed to observing on television. Yet, through each pose, and every new movement, I was physically and simultaneously living out the same experiences, mirroring the emotions and physical actions these same objects demonstrated in front of me. I was them, and they were me. Through them, I was able to experience, in my body and in my mind, whatever I directed. Bending, flexing, feeling—all were reproduced and channeled through my tiny frame as a vessel. This was real, tangible. They reflected in me a simpler form of what I felt watching other people.

As such, I was meticulous, keeping my toys in pristine condition. I generally tended to their needs, though I might occasionally drop or break one. If I did, I would cringe, as if I had been dropped or broken myself. Deliberate injury to them was inconceivable. I might as well have injured myself. Rainier, on the other hand, would mix and match the limbs of his action figures, stockpiling his toy chest like a deranged surgeon's torture chamber. I couldn't help but literally feel as though I were being pulled apart and put together, monstrously assembled with a rhinoceros head, mismatched with a petite right arm and a bulbous and muscled left. I pitied these creatures. I would try to extricate them from Rainier's macabre clutches to rehabilitate them, at least emotionally. I was rarely able to bring myself to switch their parts back because these were their new bodies. The best I could do was to help them understand that, while these were their new bodies, they were still worthy of love and respect. With that, I thought, they might have a chance to learn how to use their new selves, perhaps even surpass the potential of their previous forms.

Rivaling my love of objects and everyday things was my deep fascination with television, my near reverence for the medium. While some children were temporarily pacified by their blankie or a sippy cup filled with chocolate milk, I was instantly tranquilized by *Night Court, The Golden Girls,* anything, really, that ran on that glimmering screen. I would sit for hours on end watching cartoons. When those cartoons ended, I would grab the tattered box of VHS tapes my father had inherited from a coworker. I might put in a collection of Goofy skits. Or, if I was in the mood, something to mix up the experience, like *My Little Pony.* It didn't really matter what I was watching. The hypnotic quality of my television-induced trances was indiscriminate—the dazzling dervish of colors and sounds guaranteed I was going to pay attention, regardless of what was on the screen. Mostly what captivated me, though, was the full experience of immersion whenever a show began. As soon as my eyes caught sight of the television's flickers, I was enveloped. It was everything. Like surrendering to a dream, I became an ethereal presence within the television world. Amidst this technicolor fog, my body mirrored every touch and every movement, shifting in successive quantum leaps from one perspective to the next. Whenever bailiff Bull Shannon slapped his forehead, for instance, I felt the wisp of a slap against the same part of my own forehead. Whenever the Road Runner stuck out his tongue, I felt the sensation of my own tongue blurting out of my mouth, while the sound effect evoked rounded ash-colored bubbles. The wild contortions of Tom and Jerry echoed through me. It never occurred to me that no one else experienced television like this.

Television taught me the fine line between numbing and immersion. I could watch a new episode of a PBS special and learn to truly appreciate the vastness of the universe. Or, I could watch a rerun of *Rabbit Punch,* the Looney Tunes skit where Bugs Bunny steps into the boxing ring against "Battling McGook," the reigning champ. I could choose to watch the rerun for the sole purpose of wasting time, transporting myself away from the reality of arguing parents or my embarrassing lack of friends, and somehow appreciate at a young age how the numbness made time fizzle away as quickly as the scenes I fast-forwarded through on the

VCR. However, I also found myself stepping out of this numbness for brief moments at a time as I noticed something I never saw before during a rerun. Perhaps I would notice the discontinuity of Bugs Bunny's pink trunks, which were sporadically purple for occasional cells. I might notice the anger in the heckling of the audience around the boxing ring. Mere moments like these were enough to reengage me.

In this way, television provided me with lessons below its pixelated surface. It was intoxicating. It was educational. It was soothing. As the pixels danced on my screen, all of my thoughts and actions stopped midstream. I remember sitting on my family's brown textured carpeting with my mouth agape, my eyes fixed for hours. This unrelenting response to condensed stimuli persists throughout my life. Though as a child, there was a cold solace from being instantly drawn out of the "real" world. It felt as if there were too great a distance between me and the rest of the world—everyone, including my family. Because I didn't know how to close this gap, I too often completely let go of my grip on the external world, immersing myself even deeper into my internal world.

Books provided a similar immersive pull. While books were less voracious, novels would coax me down a staircase, word by word, into the basement of a new subreality. Within a book, particularly Choose Your Own Adventure books, I would enter a dream where the images evoked by words not only existed in my mind but would also slowly take over all of my senses. Like television, I would slip into a separate world as a phantom observer experiencing the lives of others through various perspectives. I eventually had to wean myself off the Choose Your Own Adventure genre because their intoxication lingered into reality, and I might find myself fussing over endless inane choices. If I open the car door, will I find the coiled rattlesnake that will bring my story to an abrupt end? If I answer this phone call, will the raspy voice of a stranger give me the clandestine instructions that will start me on my adventure? Despite all my hours hunched over the pages of a book in dim light and all my hours of screen time, which would throw any committee of pediatricians into a kerfuffle, I am actually quite thankful to my parents for giving me the liberty to explore these worlds. In the times that I was engaged and kept myself from drifting too far into the deep, these worlds and their inhabitants taught me English as well as the basics of communication: what to say and how to say it, plots and archetypes that ring familiar with other people—a working knowledge that allowed me to relate to others as an American and as a fellow participant in the currents of popular culture.

Still, these living layers were the whole of my reality. Within my mind were rabbit holes upon rabbit holes, which I often used as makeshift shelters to protect myself from the harsher, estranged reality of the external world. In the late 1980s, my parents could no longer afford to live in the United States, despite their best attempts. My mother worked long hours behind the counters of the local supermarket deli or bakery. Setting his alarm for 4:00 A.M. every night, my father delivered newspapers, tossing them out of the janky windows of our family's rusted gray station wagon. Once he finished with his paper route, he would change into his Pullman brown uniform and begin the labor of lifting hundreds of cardboard boxes into the scalding heat of a UPS truck, delivering them throughout South Florida. Hours later, he would return home with the thick smell of physical labor. He would shower, then change into a Maya blue polo with an amaranth-colored collar, place a snug Domino's Pizza baseball cap on his head, and rush out once again to start his pizza delivery shift. I saw him just enough to feel the weight of his labor, which echoed in me his sluggish, imprecise movements from fatigue. The anguish under his eyes hung low, and he was often too tired to force a slight smile. Watching him, I felt my muscles tighten, requiring more effort on my part to smile, to fight against my father's imprint. He was shackled by bundles of knotted muscles in his lower back; I felt the sensation of my torso steadily ratcheting forward as if quietly being forced down toward the ground by an invisible hand. His hands, roughed up from experience, were calloused but warm. Running my hands across the ridges of his palms conjured the image of dry river rocks under a soothing summer sun. The physical strain and discomfort I felt belied his confidence. In his tireless smile, I could feel the strength of conviction that can only come from a man who wants his family to thrive in a new country.

From time to time my miniature body felt the echo of a void in my parents, a sad emptiness. I would feel it when, in the corner of my eye, I recognized a blank stare, fixed on the space before them. It was a lifeless gaze of longing and distance. My mother and father had fled Nicaragua as political refugees. I regularly mirrored the tension in their eyes and throats, which would feel as though the oxygen in the room had run out, my throat suddenly grasping for air.

But no matter how they tried to make use of their education in America or how many hours they toiled, they could not avoid the inevitable bankruptcy that forced them to move back to Nicaragua, where I felt for the first time in my own body the decayed teeth and scaphoid abdomens of poverty. Seeing the dirt-caked eyes of a small emaciated child in a tiny hut next to an unpaved road, my eyes felt the rolling of pebbles and dust behind my eyelids, scraping along with every blink. I winced and squinted to protect my eyes from phantom debris and to protect myself from as much of the raw suffering as I could. The swollen bellies of children made my abdomen feel hollow and distended, much like the engorged *panzas* of men who, swearing allegiance to the code of machismo, stuffed themselves with *chicharrones* and cigars. I found this education in suffering, pride, and helplessness disturbing, then cathartic, and, eventually, fundamental in understanding gratitude and compassion. But these lessons came at a price. While living in Managua, soaked in the feeling of being thin and emptied, I craved the womb of prime-time cable television. Unexpected strikes regularly cut off our electricity, and in the rare moments when local television stations came through, they appeared in garbled, unintelligible spurts. Lacking my reprieve of choice, I quickly learned to numb myself through food. It was as if I were trying to fill all the missing conveniences of a developed country with rice, beans, tortillas, or anything else edible. The spongy consistency of a thick tortilla evoked the feeling of being swallowed by cool, white clouds billowing above my head. The warm, thick pasty consistency of refried beans provided a sensation of lethargic gray melting, which helped to drown out the visceral pain of starvation, which I witnessed all around me. Filling myself was the only way to counter hunger's haunting, gnawing echo. The stretch inside my stomach was like a ball of fur, slowly expanding and filling my insides with crimson surrounded by a thick black outline, muting the memory of what my eyes mindlessly translated into acute physical sensations. Day by day, I became heavier, more cherubic in the face, wider around the waist. If dressing was difficult before, it became even more of a chore after I put on weight. My pants became more and more uncomfortable against my skin. Running caused me immediate cramps, which stabbed my already tormented insides.

Fortunately, we moved back to the United States eighteen months later. Though our return helped alleviate some of my day-to-day concerns, I was still faced with the same feelings of isolation. These feelings of isolation were exacerbated as my brother and I continued to drift apart. Rainier had an easier time making friends with cousins, neighbors, and, to my dismay, strangers. At birthday parties and other family gatherings, my brother would effortlessly join the ranks of the other children in our age range. I, on the other hand, remained an easy target. To cement his place within the group, Rainier often instigated scuffles, which could easily erupt into an all-out fistfight between us. Because each punch felt like a swing at my brother and myself simultaneously, I pulled my punches so neither of us would get hurt. Meanwhile, Rainier practiced a near-constant hyperactivity at home and in class, wrought with the additional inability to focus for sustained periods of time. Getting diagnosed with Attention Deficit Hyperactivity Disorder embarrassed Rainier. Feeling as if he were flawed, that something was inherently wrong with him, should have brought Rainier and me closer. But, as I later came to understand, Rainier resented me because he felt our parents unfairly compared him to me, his quiet and weird brother, who, though largely disliked, at least never got into trouble. They told Rainier to act more like me, to talk more like me, to be more like me. And this, of course, only alienated us even further. I learned to accept this and, as Rainier and I grew older, the distance grew between us, too.

Throughout elementary school I was increasingly pushed outside the primary circle. Sometimes this was subtle. Perhaps I might notice that I seemed to be "it" at the start of tag more often than chance, or I might realize that "saving the best for last" is irrelevant in picking members for a team. Other times, it was less subtle. There might be real effort on the part of others to get under my skin, to tease me until I lashed out. But by this time, I was so adept at numbing that I could block everything out, refusing to even appear vulnerable to their teasing.

There certainly wasn't an absence of an emotional response,

though, especially when I was provoked enough. The bulk of my response occurred either internally or was delayed. I remember crying a few times. But I made sure to cry only at home. I recall feeling sad, obviously, about my classmates' behavior but also confused about why I experienced their insults as a form of self-hatred. Listening to their taunts, I felt as though I were provoking myself, pushing and pulling myself back into the darkest corners of my insecurities. My anguish, born from sadness and confusion, was sharp. I experienced it in my chest as a sheet of an unearthly onyx-colored material with a reflective surface. Like an aluminum can being crushed, it would forcibly crinkle and bend and buckle in on itself. The sharper it bent, the more painful it became. It was heavy in my chest, with screeching, piercing corners. Questions overflowed and poured out of me as tears. Why would a person treat another human this way? What was its function? Was it because it helped bring them closer to those they considered to be their friends? Was it because someone taught them that this is what you do to those whom you think deserve it? Such questions beckoned answers, but dark thoughts continued to slink through my mind. These thoughts came in a whisper—"they're right!"—as if hissed by a gnarled troll from the shadows. Was there something inherently *un*-good about me? And, if true, would I be stuck like this forever?

I remember asking my mother why no one seemed to like me. All I wanted was to provide them with happy moments. I wanted them to experience the same warmth and cool invigoration that I experienced when I hugged someone or saw two people embracing, which for me was the pure feeling of the number 4. My mother sat lightly next to me on the bed. The old mattress creaked, and the box spring cleared its throat. She gently rubbed my back. As the sting of confusion and frustration started to blunt, the mysterious onyx material in my chest began to relax and slowly straighten out. I peeled my face, round and red from crying, away from the pillow. "They just don't know," my mother told me. "They just don't know what they're missing." I appreciated my mother's soothing words, not because they validated my worth, but because they countered the troll-like whispers about my insecurities with an essential truth about people. We just don't know. We may have the same constituent parts and act in ways that appear, at least on the surface, similar. But we think differently enough that we may as well exist in infinitely separate worlds.

In response, I sought to obtain an external world that would more accurately reflect the greater connection I had with others below the surface, though my childhood track record in securing such a connection had mixed results. Around this same time, I was invited to a birthday party. The party was hosted at Hot Wheels, a roller-skating rink that specialized in children's birthday parties and bruised knees. My father dropped me off and said he would be back when the party was scheduled to be over, two hours later. It was one of the few real school-related birthday parties I was invited to and, in my prepubescent warped perception of time, two hours seemed like plenty of time to build lifelong friendships in between laps around the rink. I might even have time to spare. I stepped into the busy party area outside the skating rink, power-walked in the general direction of neon balloons, and immediately shook hands with the birthday boy's mother. She was courteous and pointed me in the direction of the others. I paused. There were a few familiar faces from our third-grade class but many more faces I didn't recognize. The birthday boy wore an oversized cardboard crown that pushed down his ears, jutting them outward like one of Santa's elves. I stood there and watched the kids scream and throw popcorn at each other. By their laughing and smiling, it looked like it was fun.

Eventually, as the other kids continued running around aimlessly, I approached the group. One party-goer briefly used me as a shield. While it was fun to feel the wild movements of their bodies projected on my own, I eventually tired from standing, staring, and attempting to laugh and smile when the others did. I retreated back over to the table with the adults, where I knew I would at least find lukewarm pizza and pitchers of flat soda.

Spending time with adults was usually far more interesting anyway. There was often more to experience. I could superficially comprehend what was going on, what the adults were talking about, all while picking up a fair amount of vocabulary. What I enjoyed most, however, were the secret conversations they had with their bodies. Compared to most kids, adults were far more animated, much more expressive. It was like watching television. I could hear, see, and feel emotions and thoughts that were heavy and dark, full of contours like aged cherry wood, yet still playful. The wrinkles on their faces not only piqued my interest, but also made it easier to follow along with the expressions on their mouths. I could feel the exhaustion from traffic and driving the kids around and the tiny smiles that flashed unexpectedly after a comment that did not clearly sound like a joke. A flash of teeth, pouted lips, squinting eyes, uncomfortable silence. I felt them all equally, and I loved every minute of it.

I made a game of seeing if I could guess who was going to speak next. There were often distinct changes in posture and expressions that communicated to the others that it was someone else's turn to talk. With each turn, my predictions became more accurate. By sitting around adults more and more, darting my eyes from person to person, I learned valuable lessons in helping to kindle specific emotions in others so I could better communicate my internal thoughts to the rest of the world. If I failed to do this, my message would often fizzle away into the atmosphere.

Or worse, if I accidentally kindled the wrong emotion, I might find my message turned completely on its head. The easiest way to keep a conversation going, I realized, was to look a person in the eyes as they talk, then relay something similar about you or your life. The other person usually does the same in return, and the pattern repeats itself. I was learning to bridge the distance between me and other people by mirroring them, and perhaps the most important lesson I picked up was how to maintain a conversation long enough to form a connection—however slight—with another person.

What was said didn't necessarily matter. Sometimes the words didn't translate accurately what was going on beneath the surface anyway. Once I realized this, I was free to focus on my body's reflection of what people were actually trying to communicate either to me or to another person. Listening beyond the words to comprehend their true intention allowed me to clarify what others *wanted* to communicate rather than what I *assumed* they were communicating. This required me to pay close and careful attention to what the other person was conjuring in me, to get more in tune with how my body was automatically translating and reflecting information, and then let the other person know with clarity that I understood.

I continued this practice through high school, but I was not prepared for the suffocating level of aggression that oozed out of my new school's linoleum-floored halls and gravel parking lots. Every day the bilious angst from hundreds of other disadvantaged teenagers would angrily agitate and froth. Making my way through crowded hallways between periods, I could feel the tension building below my skin. Of course, the anxieties that come with typical adolescence crawled under *everyone's* skin. But beneath these skittering centipedes there was also the slow, rolling, rumble of anger. During the rush between classes, countless faces revealed the tide of frustrations and a current of mindless primal hunger gliding beneath the surface. Bumping into shoulders, catching onto other's backpacks, you never knew when these emotions would explode into physical violence or confrontation. You would suddenly find yourself shoved to the ground, punched in the face, or tripped, your feet pulled out from under you.

When the blue and yellow trilling of the school bell rang for the end of each school day, I reflexively felt the joy of escape just before the dread of the final obstacle: surviving the school bus, an unsupervised cheddar-colored chamber of rage. As I walked through the unpaved lot behind the school, the crunching of gravel under my steps sent soft gray lashes flittering through my feet. The best I could do to survive the final part of my day was either to find a small tribe of allies to band with or become invisible. If you didn't run to your bus early enough, before the rush of other raging passengers, you would inevitably find yourself caught in a brawl that broke out for one reason or another. I was synesthetically flung into every fight I saw. The reflected anger seeped through my core. Trying to avoid a tussle between two sweaty teenagers entangled like cobras, I felt knotted bright scarlet threads dragging crimson and dark maroon coils through my arms and legs, echoing deep growls of feral aggression. Like a bystander caught in the line of fire, the sensation of punches ricocheted through my visual system and found home in my sinews. The sensations of small repeated shoves, being pushed backwards, resonated across my chest, though I wasn't actually fighting anyone. Landing on little jagged rocks, pressing into my palms and lower back, reverberated through me. Clouds of earthly and phantom dust scratched at my eyes.

Clawing was the worst. I once found myself caught near two screaming girls. One had a series of gold bracelets and thick gold earrings that jangled as she threw the first punches. The other girl lunged with her right hand, her long electric orange fingernails angled like the claws of a tiger. Then each took her left hand and clasped on to a handful of hair. From the top of my head, I felt my hair nearly being uprooted out of my scalp, as the two girls clung and pulled at each other's hair in an effort to tear it off. They spun each other in circles. Their right hands remained free to swing around and claw downward into the other girl's face, trying to cut into her skin. In rapid succession, streaks of what felt like pointed arrowheads raked against the right side of my face. To drown out this raking sensation, I dug the palm of my right hand into my face. I jumped into the bus and plopped into a seat by the window facing away from the brawl. Through the warped dust-covered window, I could see the stampede of students running toward what had begun to look like the crowd of a back-alley cockfight. The burn of the tattered vinyl seat, heated by the relentless Miami sun, was a welcome distraction.

With each day of high school, I felt the distance between myself and others grow wider. Like everyone else my age, I was desperate to understand my body, how it responded to hormones and physical touch. But unlike everyone else, I seemed to be the only one entirely burdened by it. I was considered weak in the hallways, squeamish and overly sensitive, strange in the classroom. I felt myself silently repelled by others and, thus mirrored, repelled myself even further. It became harder to hold myself in any esteem when the message of worthlessness reverberated so loudly, so physically, through the halls.

I also didn't walk like the other males. Nor did I talk like them. I didn't pick fights and I avoided rather than sought out conflict. I didn't have a girlfriend and I didn't talk about "liking" anyone. The verdict from the others was that I was gay. But I wasn't gay. Was I gay? I couldn't be gay. What if I were? The distance between me and anyone else was still too great to conjure any feelings of attraction toward anyone else.

Fortunately, learning brought me joy. It was only after I had survived the end of my second year of high school that academic performance finally became a vicarious indicator of social value. Of course, being a "nerd" had no celebrated street value. But my performance in class had excelled enough that the respect for my academic and intellectual ability slowly began to permeate outside the handful of advanced placement and honors classes.

Respect could not replace friendship, however. I needed actual friends, somewhere I could act as an extension of others and they could act as an extension of me. The fundamental issue, however, was how I tried to connect with others. Interactions in real life were nowhere near as straightforward as advertised on television. You couldn't simply say what was on your mind or ignore what the other person was saying or doing. I had to dig deeper. I began to use my facility with learning to more deliberately study others, then apply the knowledge acquired from my observations toward interacting with my classmates through the subtle clues they unknowingly communicated.

A classmate once remarked to me in the cafeteria that the food sucked. The left corner of her mouth angled downward, and her right nostril pulled upwards. She was serious, but because her left eyebrow was raised slightly, maybe she didn't find the food completely unappetizing and really just wanted something else to eat. The pause in between her words, coupled with her raised eyebrow, probably meant that she was waiting for me to agree or possibly offer a solution. Though I still needed to figure out if this were true or just speculation, I knew the best way to test a hypothesis is to experiment. "I know, right?" I attempted. "Want to get something from one of the lunch trucks outside?" The snark melted from her face as I felt growing anticipation mirrored on my face. She smiled, "Ohmaigad, bro, you totally read my mind. Let's go."

Sometimes I was spot on with deciphering the hidden question, and my responses would be well-received. It certainly was not telepathy, but it was helpful, although assuming and being wrong was often more dangerous than simply asking follow-up questions or replying with a statement or action designed to gather more information. I soon learned that I had to express the words with just the right inflection and the right facial expression and body posture in order to acknowledge that I first understood what another person was asking, then respond appropriately. "Your haircut is amazing!" stated with a wide mouth, raised eyebrows, and two thumbs up could be interpreted as sarcastic, compared to the more genuine and subdued "I like your new haircut."

Despite the iterations, however, I kept finding instances where identifying the underlying question was not enough. In some cases, I might identify that the person was feeling exasperated or depressed. But when a classmate simply wanted to go unnoticed, addressing with them what I was picking up emotionally was sometimes the worst possible thing I could do. Classmates might not want to talk about their feelings. They might feel self-conscious. "Oh god, is it that obvious…?" They might feel vulnerable, particularly if they hadn't consciously tried to communicate their emotions. "Dude, what the hell are you talking about?" In these instances, it was either irritating, completely wrong, awkwardly misplaced, or all of the above. I fumbled through interactions like this until I was able to reach a steady state where I could say that I had a handful of acquaintances at school and was slowly brought in closer to multiple groups. I gained confidence in my interactions with others and began to shed some of my invisibility. It became less of a gamble to speak in my own voice, in my own way.

There were still lingering sensations of what some might call social dysphoria. But at least I had begun to get the hang of unspoken communication, the small step I had been desperately trying to take. And then, the summer before my junior year of high school, I unexpectedly met the first of who would become my three best friends from high school. Alex and I were sitting a row away from each other on a school bus chartered for an SAT prep course at the local community college. We struck up a casual conversation about our upcoming classes and, as these things often do, the topic invariably devolved into a full-blown geek-out about anime. Because the conversation between me and Alex required little effort, it stood out in my mind as profound. Every morning, on our way to the community college, and every afternoon, on our way home, Alex and I delved deeper into our geekdom. It was glorious. This was clearly a friend.

Once the summer program was over, I hoped to continue my friendship with Alex during the regular school year. We happened to be taking a few advanced placement classes together, so I figured it would be simple. I noticed that he would have lunch with two other guys out in the middle of our high school parking lot under the shade of a tree. Emboldened by our previous conversations, I walked over to say hello and have lunch with Lee, Steve, and Alex. It took no more than a few seconds to feel how alive the mirth was in this group—warm yet cool, like standing in a misty oasis. Every statement and exaggerated comedic gesture echoed a gentle glory of relief, of mischievous confidence, followed by the other two laughing or chuckling. I was tickled by the sensation of laughter and enraptured by the feeling of my own laughter, resonating in harmony with the group. I was also engaged by the layered nuances of their jokes, often dipping into a collective well of shared memories. They were often tied to television, science fiction, comic books, or anime.

When I came back the next day, however, no one was at the tree. I extended my neck to gain a greater vantage point of the parking lot, scanning across the noisy static of several throngs of other students. I got a glimpse of Lee's face in the distance behind a lunch truck. I stuck my thumbs under my frayed black backpack straps and bee-lined in their direction. Though they kept on eating in different parts of the parking lot each day, I always found them. Spending lunch with them was the best part of high school. The more time I spent with them, the more I was able to acclimate to their slapstick physical humor and see how to deliver a story in a way that had a high probability of making others laugh. It was also a space where I could practice this new exciting element of the conversation game. I was utterly clumsy in my delivery, but after many failed attempts and, after what felt like hours and hours of observation and practice, I steadily became better at it.

Over the next couple of years, we continued to grow closer and, to this day, consider ourselves spirit-bound brothers. Though it was ten years into our friendship before the three of them finally confessed over drinks one night that the reason why I had to keep looking for them during lunch for the first months of getting to know them was because they were actively trying to avoid me. Unbeknownst to me, they had nicknamed me "the Joke Killer" and didn't want to have anything to do with me. But I was so engrossed in the laughter and the subject matter of our conversations that I never even noticed. Even if I had, though, I wouldn't have cared and would have continued to look for them anyway. I'm so glad they stuck through my awkwardness for as long as they did. I'm equally glad that I was as persistent as I was. Otherwise, I would've missed out on so many wonderful memories. Through interacting with them, I was able to appreciate the importance of humor and, in the process, learn how to make others laugh. The ability to incite smiles, amusement, and comedy in others was one of the most valuable lessons I learned. Shared laughter, not unlike a good hug, is an embracing, radiant joy that reverberates warmth through me in a beautiful and indulgent continuous loop. The joy I create in myself in doing so radiates to others, and their joy reflects in me. When this happens, I experience the rawness of joy and laughter at the center of my head and in the pit of my stomach, what I can only describe as the fluttering and blossoming of iridescent wildflowers.

My friendship with Lee, Steve, and Alex was the beginning of a new phase in connecting with others. Though, as the last year of high school came to an end and, after months of figuring my way through college applications, financial aid documents, and more with the same tenacity I needed to figure my way through high school, I was both saddened and excited to move on to Cornell University, driven by the hope that by persistently (and thoughtfully) pushing onward I would continue to grow closer to other people.

This new environment allowed me chances to study myself and others at all hours. With my senses and all of this exposure, I developed at some point a more reliable system in connecting with others through recognizable patterns, including cadence and tempo of speech, more pronounced facial expressions, more accentuated body movements, and more appropriate responses. This was most evident in groups, where I could watch how people interacted with each other, where I could observe conversation without having to participate. Juggling intense observation and engaged conversation was to me like bird watching and gardening simultaneously: both lovely activities, but I ultimately ended up doing one better than the other while doing both terribly.

If I could estimate that the likelihood of my interaction with the group would be brief, then I was more inclined to take risks.

There was little downside, I figured, because I wasn't likely to see them again anytime soon. In these scenarios, I was considered an extrovert. If, on the other hand, I were likely to spend a lot of time with a group of people, I usually spent the first day or so observing, interpreting. Once I felt comfortable with my observations, I could burst into streams of engaging conversation, pointed questions, and tongue-in-cheek remarks, which to my delight summoned chortles, chuckles, and guffaws. In these instances, I was considered the introvert who miraculously became an extrovert, a likable and entertaining part of the group. A hiccup in this process, however, occurred whenever someone new unexpectedly joined the group. People's behaviors, I noticed, shifted dramatically with the addition of a new person. I often had to reassess the dynamic, then start over with more intense observation before reengaging, a disorienting but necessary process.

Often, after spending more and more time with one person, I would realize that I was inadvertently mimicking his or her behavior—placing a pencil on my lower lip, for instance, or standing with arms akimbo, speaking exuberantly, or appropriating phrases someone liked to use. This wasn't the shadowing of echopraxia or echolalia often found in people with tic disorders or Tourette's syndrome, nor was it the deliberate use of another's expressions out of admiration or mental availability. Rather, my mirroring seemed like a cloning process, a living clay effigy lathed over my bones, that occurred reliably and involuntarily. Interestingly, after spending a few more days together with the person I was mimicking, my mirroring would suddenly stop, as if I no longer were susceptible or acutely attuned to their behavior. Though having felt how their faces and bodies moved, I could tap back into mimicked behaviors and speech patterns when I needed to, like a subconscious Swiss Army Knife. Trying to mediate a childish argument, for instance, my brain might reconjure my dorm's resident advisor, employing her deliberate pauses, punctuated with light-hearted remarks that ended on a down-flection in vocal pitch and a blank stare to put the matter to rest. If I were asked a serious question during an otherwise light conversation, I would tilt my head up and to the right while wearing a sturgeon-faced grimace, squinting and blinking to sarcastically ease the transition—just as my roommate Kensuke often did. This pattern continued through my freshman year—greeting, observing, learning, feeling, becoming, incorporating.

Aside from a small handful of friends, however, I tried to avoid interactions unless they were absolutely necessary. I would move and speak specifically to blend in with others, easily forgotten. Video games, like television in my childhood, offered me an easy escape. Massively multiplayer online role-playing games, to be exact. I spent hours upon hours in front of the computer screen immersed as my avatar. I could feel his movements, his kicks and punches, the subtle sweeping of his hair. I was in complete control over this world. The lights, sounds, and colors that flashed, exploded, and radiated were far easier to understand than the complexities of real life. This made it harder and harder to pull away, and I continued to hide even further from others.

When I did interact with people, there were certainly times when my behavior was considered offensive or off in some way. But I learned to steadily gauge my sensitivity to other people's actions and behaviors. I had to be sensitive enough to not give these kinds of off-putting impressions while at the same time not act so sensitive that I would be constantly paralyzed in speech and action or so accommodating that I would become more of a convenience than a companion. I came to acknowledge this balance the hard way. My sensitivity to other people's reactions to everything I said or did steadily amplified. This was particularly true with Cristina, my girlfriend at the time.

We first met at Cornell during a summer program but ended up going to different universities. She and I maintained a long-distance relationship with periodic visits and frequent phone calls, text messages, and AIM (America Online Instant Messenger). She was my "first kiss." After our initial clumsy rendezvous when we were seventeen years old, I grew to love her. I hesitated at her first advances, but I still wanted to be around her. She was vivacious, unforgiving. Feeling the weight and certitude of her movements was anchoring, alluring. The first kiss began on my left shoulder without my consent. It soaked through my skin and into my bloodstream, spreading its intoxicating warmth and coolness, undoing every knot it found under my skin. My body trembled with relief. Her kiss trailed up along my neck and ended on my lips. I lost myself. Whether days, weeks, or months thereafter, I don't recall the exact moment when "interest" became "attraction," or when "attraction" slipped into "love." I don't know if it's a moment that can be found in the synapse between two neurons. But at some point, my brain's emotional and reward networks conspired to carry me across a threshold, nearly eliminating the line between me and her.

Rather than celebrate having found a connection similar to what I had been searching for my entire life, a new concern was exposed. Without a boundary between me and another person, I cohabited two bodies. Overconcern. Identification. Enmeshment. Dependence. Stockholm syndrome. There are multiple intersecting terms used in psychology and psychotherapy to describe the pathologic condition of a person irrationally believing that he or she has assimilated, or is one, with another person. The delusion can be so great that a person sometimes finds himself or herself emotionally distraught over anything that might compromise the welfare of the other. In retrospect, I instead experienced a phenomenon of *eros,* romantic love, painted with hues of all these concepts. The more time I spent with Cristina, the more she became a natural extension of my nervous system, of me. Our *eros* was not a feeling of expansion or completion most people in love experience, or hope to experience. It felt to me simply like a loss, that I was less myself. Like my fights with Rainier, the potential of harming Cristina was just as threatening as the potential of harming whatever part of me that was still left. To offend her was to offend myself. Whenever I unintentionally and inevitably did offend her, it was usually a result of my own awkward interpretation of something she said or did or my failure to communicate my thoughts and feelings accurately. This was amplified through the same emotional volatility that I had initially found so charming, so intoxicating. With every misstep, a bomb would go off on the vast minefield of our interactions.

Keeping Cristina happy was how I preserved bits of happiness. Her attempts at making me happy were only successful if she, too, found enjoyment in such moments. Because of this, I slowly faded out of the equation, and our partnership, though no fault of her own, devolved into a relationship between Cristina and an opaque reflection of Cristina herself. I sought to avoid conflict. But the distress she manifested was imprinted on my body and I, in turn, felt only the distress I had caused her, experiencing it as echoes of shame, guilt, and suffering.

And then, one day, a thought slithered into my patchwork shell: *Who am I?* When I looked in the mirror, the question weighed heavily in my lower abdomen. Was I the reflection or the person reflecting back? I could make out only reflection after reflection, a funhouse mirror, without beginning or end, lined with illusions of an identifiable, independent self. Or, perhaps I had become one of Rainier's macabre Frankensteinian creatures, a chimeric figure pieced together from the various constituents of human remains.

I needed help. For most of my life, I had been able to reason through earlier, gentler versions of these thoughts and escape more or less unscathed, intact. But as I continued to sink into my reverse solipsism, I realized I couldn't do this alone. With Cornell's on-campus mental health system, I was able to see a therapist rather quickly. We only met a few times, but it was enough to help pick out questions I could chew on and then work through.

The online group of friends I had grown close to by then was an important complement to continuing this process. Disembodied chunks of text from gaming avatars were essential to creating a safe space where I could focus on myself and air out my thoughts and feelings devoid of any perceived mirrors in judgment or self-recrimination. By communicating through text only, I learned to negotiate through the structures that made up my mind. Making some errors in interpreting others was more acceptable in this medium. And, in time, I came to appreciate that the people and experiences that shape us do not *define* us. They *refine* us, as imprints that first leave wounds and then tomorrow turn into scars and callouses that allow us to climb higher.

But it would take many more years before I understood this in full.

CHAPTER 2

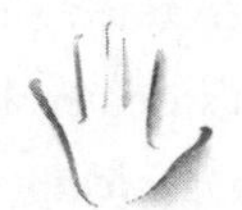

Purple White Carrot Montana Dust

The Kayapo, an indigenous tribe in a remote part of the Amazon rainforest, interact with the world in a fascinating way. They interpret emotional and physical injury—no matter how painful—as an absolutely hilarious experience, a tremendous source of humor. They express pain through laughter. A stubbed toe or an excruciating bite from an Hormiga Veinticuatro (or bullet ant), it didn't matter. They hold dominion over their environment and how it affects them.

Canoeing down the river one day, I watched an elderly Kayapo woman making her way up the steep river bed, a heavy basket full of Brazil nuts slung around her back. Near the top of the ledge, she lost her footing and tumbled twenty feet down the river bed. Looking on in terror, I paddled frantically toward her. But as I neared, I heard her laughing. She was dusting herself off, giving out a lively laugh while the adults and children around her joined in, their communal laughter echoing up and down the river bed.

In a poetic reversal, the Kayapo, whom I worked with during my junior year in college as part of an ecological research trip, also translate their emotions into physical pain. A mother mourning the loss of her child, for instance, expressed her sorrow by slicing two-inch crevices in her scalp with a machete. When I met her, it had been years since her son's death, but her grief was still apparent. Her head was covered with long, ridged scars, which felt like dried wild reeds lying against my own head. A dull ache radiated and draped over my head like a funeral veil. The woman's sorrow echoed and resonated in sync with the onyx shrapnel in my chest, yelping as it buckled into itself. Meanwhile, when Kayapo fall in love, they scratch the shoulder of their partner, the object of their affection. The deeper the love, the Kayapo explained, the deeper the scratch. Throughout the village, thin scars, which I wore on my shoulders like phantom epaulets, marked every man and woman who ever loved someone or ever once felt love.

How the Kayapo use their bodies to express pain and joy resonated with me. It was as if they had turned their bodies inside out, displaying for all the world to see what they were feeling internally, a heartstring to be tugged. Their scratches and scars weren't clues to some mysterious inner condition; they were the feelings themselves, made manifest, visible, and as such, easily identifiable, not unlike tears from the eyes. Because of this, it was impossible not to empathize with the Kayapo, to recognize immediately the emotions roiling beneath the surface. I didn't have to search out specific cues to figure out what someone was truly feeling or what they were trying to tell me. My experience with the Kayapo helped me realize how empathy—noticing, interpreting, and experiencing—works and how I personally interpreted and experienced the world in a heightened, nearly constant state of empathy.

Observing how the Kayapo managed physical sensations and

emotions simultaneously, as if they were one and the same, was easily one of the most inspiring experiences of the entire expedition. And following this trip, I decided to become a healer, though I approached the idea with trepidation. Do I become a healer by pursuing medicine? I hesitated. The bittersweet experience of working with patients was likely to be challenging—more challenging than the typical difficulties associated with becoming a doctor in the first place. I wasn't yet aware of synesthesia, but I was very much aware of my own internal experience and I knew that to be a *healer* would require taking on the heaviness of others—rejoicing and bearing the boon and burden of reality together. As the Kayapo understood, there is always pain. Pain, illness, death, and suffering, all of which are inevitable. But it is our reaction to such things that matters most. "I could do this," I told myself. I might not be able to do it the conventional way. But I could dedicate every fiber and neuron in me to become a physician, a healer, and do it in a way my body and mind can handle.

In my final year at Cornell, I applied to medical school, eventually getting into the University of Miami Miller School of Medicine after a brief stint on the school's waitlist. The city I had left was harkening me back home. Returning to Miami gave me the opportunity to spend more time with my family, especially my brother and sister. I could learn and help care for patients in the epitome of cultural diversity, an environment where I would be able to apply myself in full force.

Cristina decided to come with me, leaving behind her friends and family who, along with my own friends and family, celebrated our decision to move in together as the final step before our inevitable engagement. Living together and creating a home for each other certainly felt like an engagement. Before our move, we talked in depth about how challenging it would be for her to live in a new city where I would be spending much of my time in school, or in the hospital, or studying.

The more time I spent with Cristina, the more entwined my senses *and* emotions became with hers. Cristina's frustrations with Miami grew as she felt more detached from her college friends. Her most stable emotion was fear, a constant fear of abandonment, which took form in me as a cold, damp tangle of dark indigo and lavender clumps from which strands of blue and white fibers protruded. Because it was getting harder to disentangle myself from Cristina's overwhelming emotions, I knew I needed to end the relationship. And, around this same time, I was starting to become more and more attracted to men, an attraction I couldn't rationally explain away or ignore. I needed to escape the relationship if only to save myself.

I joined a group of medical students on a medical trip to Gujarat, India, during the winter break. Just as my trip to the Amazon helped me understand different ways to access empathy, my time in India was invaluable. Not only did it give me some much-needed physical distance between me and Cristina in order to figure out how I felt about the relationship—what I needed and wanted, instead of getting tangled in what I physically felt Cristina needed and wanted—the trip also inspired my first breakthrough about how and why I experience the world the way I do.

One night in Radhanpur, a bunch of us stayed up chatting with local volunteers, drinking hot chai. The topic of conversation turned to meditation. Most of the group acknowledged its benefits to physical and mental health. Someone pointed out how meditative states can help you appreciate the difference between the "little self," one's own perception of the self, and the "big self," the perception of the self and other selves as a single connected entity. I recall this idea resonating with me right away and even more so a few minutes later when another medical student mentioned a group of people who seem to have an easier time attaining these meditative states. This group, he told me, also has blended senses, recognizing colors in letters or sounds. I remember wondering why he would bother mentioning something so ordinary. Everyone has that, I thought. Later that night, though, I took him aside. "What makes this group of people who see colors in letters and so on special?" I asked. "Isn't that the case for everyone?" He looked at me slightly perplexed and squinted in disbelief before saying, "What? No. Not at all."

I almost didn't believe him. How did I miss this? Why hadn't I grasped such a distinct difference in perception compared to others, especially at this magnitude? Fractured pieces from my life suddenly began to fit together. I had spent a lifetime swatting away at these awkward fragments, these clues, as if they were nuisances, nagging reminders that I was different or strange. But I had never appreciated them as opportunities to ask, "Why am I different?". This may have been partly due to avoid the trauma of coming face-to-face with some of the more painful memories and accidentally reliving them.

Learning about synesthesia made me realize everyone experiences irreconcilably different realities, constructed from entirely different base materials. Of course, I always knew that we were all individuals, that we were all "unique." But uncovering these invisible differences revealed more than simple variations between people. Reeling, I imagined a group of people in a dark room, all seated in a circle around a great white bison illuminated by a single overhead spotlight. Some people look at the bison through binoculars. Others look at it through sunglasses, some through stained-glass lenses, still others through X-ray glasses. How would all the people in the dark room know they were all staring at the same bison? What would they think when someone's observations about the bison differed from their own perception of it? Would they seek out the person whose observations most closely resembled their own?

If I had wrongly assumed all along that we all see the world through the same glasses, then what else did I have wrong about other people and their perceptions? What did other people perceive and comprehend that I couldn't? Conversely, what did I perceive and comprehend that others couldn't?

In his 1884 novella, *Flatland,* Edwin Abbot tells the story of a two-dimensional world inhabited by flat geometric shapes. One day a sphere visits an unsuspecting square, who cannot comprehend the sphere because of its three-dimensionality. To help the square understand, the sphere raises the square into the three-dimensional Spaceland above. Seeing his world like this for the first time, the square is in awe. With his new perspective, the square suggests to the sphere that a fourth dimension must also be possible. The sphere, in shock and disbelief, shames the square and casts him back into Flatland. The square fervently tries to explain to the other two-dimensional shapes what he has seen but is instead imprisoned for seven years and writes his story in a book called *Flatland* hoping that subsequent generations of shapes will be able to see more than just their two dimensions.

I wondered if I should have been describing, dissecting, and cataloging my sensory experiences all along like the square, if only to compare notes with others. Would recording or sharing that information be useful to anyone else, or was it more likely to be utterly useless, a frivolous act of self-indulgence? Who would even believe me? What objective evidence could I possibly share? From the scientific perspective, such subjective reporting would probably be considered pseudoscientific drivel. Classified as qualitative and phenomenological at best. At worst? I didn't know for sure; perhaps denounced as schizotypal ravings at the edge of full-blown schizophrenia or ridiculous self-contrived lore from the fantastical introspective land of pixies and fairies. My heartbeat quickened at the thought of being shunned or cast out of the medical and scientific community I had only recently decided to join. I needed to confirm what synesthesia was, to understand it, to explore it—scientifically and personally.

During the remainder of the trip, I was flung back and forth between such questions and intense moments of clinical care in local hospitals and community centers where I was working and learning as a medical student. Because I was still in the early stages of my training, everything I experienced was raw and unprocessed—as a student and as a synesthete.

Regardless, I delighted in questioning the validity of my external and internal perception. I questioned the existence of synesthesia and its implications. I began to question certainty while learning to get comfortable with uncertainty. I wanted to discover what else about my internal world was alien to others and if there were any significance to these differences. Were my peculiar percepts nothing more than an inconsequential hereditary quirk, like a detached earlobe? Or, did I have a *condition*? Did I need to be loosened up or screwed down? Had I been in need of some form of *treatment*? If I trusted that a credible scientist had *quantified* and declared synesthesia as a legitimate and observable neurologic phenomenon, then maybe this meant that this was just how I was wired. If this was how I was wired, then I had no problem allowing myself to be a synesthete, though I remained a closet-synesthete. This was a part of me. Allowing myself to be a synesthete would also mean allowing me to be *me*. And, if synesthesia were a part of me, then maybe so was my attraction to men. Maybe that was how I was wired, too. Allowing myself to be attracted to men meant allowing me to be more of me. This was a precious long-awaited permission slip for me and by me to be *me*—a self-declared, self-actualized, and undeniable rite of permission. For the first time, I could be more of who I am, not out of convenience for others, but because it was what *I* wanted.

In a more practical sense, the revelation that I had synesthesia was like discovering a tool that had been in my possession all along, though I had only recently discovered it. It was clunky but vaguely familiar. With it, I could begin to acknowledge and explore new observations, to bang and clang around in my own skin. I began to experiment with changing perspectives on my own internal workings while simultaneously attempting to see more clearly from the perspective of *the other,* individually and collectively. My clinical experiences let me understand firsthand that it wasn't going to be easy or even predictable. I had to learn how to hone my synesthesia, to master it, and also, perhaps, to repurpose it as a healer.

Attempting to switch between the "little self" and the "big self" was a more subtle and far more complex juggling act than I ever thought possible. More physically poignant reflections, though, always caught me by surprise, as if clutching me by the nape of my neck, declaring with calm authority, "Do not look away. Do not blink. Be here. See this. Learn." I began to witness patients' suffering with intention. I did not want their anguish to be in vain.

And so, I came into the practice of being still, to run every movement, every second, through the apparatus of my mind. Processing. Recording. Applying. Processing. Recording. Applying. Because I wasn't yet aware of my mirror-touch synesthesia as a separate facet of my synesthesia, I assumed that exposure to difficult experiences and the foreign feelings of discomfort they imprinted on me was simply a part of my clinical training. If the other medical students could endure it, then so could I.

Examining emaciated men caked with dirt, I was embedded with their scaphoid abdomen and sunken intercostal spaces. A man riddled with von Recklinghausen disease (now known as neurofibromatosis) covered my body with fleshy, bulbous, and sagging masses, the condition's trademark plexiform neurofibromas. One mass, which slumped over his entire left eye and face, made me feel as if the right side of my face had been inflated like a balloon then allowed to deflate, my body's excess skin slipping down well below my chin. Another man, this one with tetanus, lay in a cot with his back arched in an almost permanent backward-arching posture with a fixed grimace on his face. His legs were outstretched, his feet curled like tree roots, his hands clutched toward his chest. I felt encased, plasterized, mummified in my own muscles, ready to snap in two.

Young women lined up in rows up and down the sweltering "Labour & Delivery" wing, a kind of half-ward, half-operating room. The two obstetricians assigned there shuttled between the women, all of them bathed in a bubbling geyser of anxiety, dread, exasperation, desperation, sweat, and pain. My abdomen felt like a large leather ball. As I watched the obstetricians perform an episiotomy, slicing their surgical scissors diagonally across a woman's flesh, I felt my pelvic diaphragm stretching and nearly shred, then following the clamping and stabbing of sutures, stitched back together after the delivery. One of the obstetricians celebrated his tenth delivery in the hour by attempting to free-throw a placenta into a metal tub a few feet away. He missed. As the placenta splattered on the ground, I felt as if I, too, had splattered against the linoleum, crumpling flat and lifeless, burst with streaks of bloody fluid, quickly cooling to floor temperature. I felt desecrated. I felt powerless. Yet no one seemed to notice or care, not even the woman who had just given birth and who was now enraptured in maternal bliss as she held her baby against her chest, smiling. Before I could register in full what was happening to me, an elderly woman in a burgundy sari waddled over with a bucket of water, placed the placenta in the metal tub, and began to mop up the blood, which had already begun to dry into a dark maroon color on the linoleum tiles.

I tried to process these new physical sensations, attending to the nuances of each feeling and focusing with an intense concentration on my consequent emotions and physical responses to them. I was determined. These new experiences were simply part of the job and thus part of my training. They were a strain that I had to get used to shouldering. After all, I had to build the emotional vitality and the physical stamina to stand for over twenty-four hours without rest, and that might be just one of three of those shifts in a week, twelve in a month, over one hundred in a year. If I couldn't handle this now, then I had no chance of making it through to the end of my medical training.

At the same time, I witnessed great moments of tenderness during my time in India. When I watched mothers and fathers hold their children close or grown children leading grandmothers and grandfathers gently across dirt paths, I felt the physical touch of care for another while also feeling the touch of being cared for. I felt warmth and coolness, simultaneously, as though I were embracing myself, holding within me gentleness and strength. I observed same-sex friends holding hands without fear or harassment or reprimand. This reflection felt natural, liberating. But it opened a wide sense of longing. I wanted the phantom touch from them to be real in me. Bathing in its mix of relief and yearning, I rinsed away any lingering hesitation and repulsion I might have had about my attraction to men—a repulsion steeped in the fog of my own culture, which was embedded with homophobia, machismo, misogyny, guilt, and shame. These sights were as purifying as they were edifying.

I left India with a head full of quixotic thoughts, grounded though by the weight of my experiences. But I began to have second thoughts about ending my relationship with Cristina simply because I didn't want to cause her, or myself, any harm. Instead, I continued to give more and more of myself until I nearly gave myself completely away. Reserving nothing, I emptied myself into her until I was no longer me; I was simply her. There was nothing left to give. No matter how painful, no matter what I told myself later, I was going to end the relationship. I spent a day writing a script for myself, outlining what I needed to say.

I came back to our apartment early in the evening. We sat at the foot of our bed. My heart kicked and flailed. My body writhed in an attempt to reject the words coming out of my mouth. It was painful, like an incision, or an amputation. As my previously rehearsed words came out in measured and deliberate tones, I quietly apologized to my body. The onyx shrapnel let out a shrill cry as it crumpled in on itself.

Following our breakup, Cristina and I met three more times. Twice in semipublic spaces, so I would not be utterly consumed by the reflection of her pain. Our final face-to-face encounter was in our apartment before Cristina left Miami. I felt the softened sparks of her anger, the muffled physical echoing of her resentment. Our connection had been severed. My body was raw with open wounds and, later, as I started to heal, absorbed in a general numbness. But this numbness, I knew, was all me. It was mine and mine alone.

I retreated to school, finding solace in medicine. Each biological mechanism, each name for a previously unknown substructure, relaid a foundation for understanding the world. Why we eat. Why we breathe. How we live and how we die. *Why* blurred with *how.* I found comfort in the certainty that came with classifying every millimeter of the human anatomy. I seriously considered becoming a surgeon. To be able to shape the physical exterior that influenced the internal world of another person—which in turn influenced mine—was appealing.

I joined the ranks of medical students interested in providing medical aid to the people of Haiti, a country ethnographically wedded to South Florida. Our organization worked closely with Haiti's Ministry of Health. Unlike other groups, which arrived in foreign countries doling out limited medical supplies and rudimentary procedures, we joined Haitian public health efforts in rebuilding the country's health infrastructure. In Haiti, I was reminded once again of the all the sensations I encountered in Nicaragua, India, and Brazil—distended abdomens, the push of bone against the tented skin of hips and cheeks, the sharp smell of poverty that switched in my nostrils from brown to a faint powdered blue laced with a curdled black.

On a trip focused on surgical care, we traveled with plastic surgeons. I awed at how they were able to graft thin swatches of skin onto burn victims with the diligence of a human sewing machine. I saw lines of patients with keloids—lumps and bumps of scar tissue, some as large as shrink-wrapped cane toads. My body took on a verrucose *Cucurbita* skin reminiscent of the plexiform neurofibromas I donned once before in India; tight fleshy dollops plopped onto ears, necks, arms, elbows, and knees. Once, while examining a woman with a thyroid goiter large enough to encase most of her neck, I started to feel like a frog bubbling its neck. I struggled to swallow. Another man, after taking off a tattered blue baseball cap, pointed at a golf ball-sized lump below the skin near the top of his head. It looked as if it were about to burst. Immediately, I felt the bump on my own head, as if I were about to sprout the horn of a unicorn. The feeling was so intense and unexpected, it nearly caused me physical pain, a rare but distinct expression I later learned of my mirror touch. I did my best not to flinch, to be there for the patient, to learn how to care for him.

Traveling between Miami and Haiti may have also helped me to come out. Perhaps because I was slowly starting to get more comfortable with myself. I had grown up in a household where "gay" was equated with physical disease, mental illness, and moral denigration. Every time I was attracted to a man growing up, I recalled stories my father told me about how he and his friends would beat up and throw rocks at anyone even accused of being gay.

After my relationship with Cristina, I felt it was my responsibility to better understand myself before engaging with another person. It was logical and civil. But, most of all, it represented a generous and necessary act of compassion toward myself. With only one out lesbian in our medical school class, I did not have gay friends I was close enough to, to discuss such things. To regroup and recenter, I returned once again to the virtual world of gaming. I came home to friends and acquaintances shrouded in pixels but behind their typed words possessing every shade of human. To me, this was not an escape. It let me simplify my thoughts and my feelings until I was comfortable enough to come out.

Late one night, while playing with a regular gamer, she and I started talking about our lives offline. We talked about what we hoped for and what we feared. She opened up about her difficulty finding a job, how she was still living in her mother's home. She also told me about the men she dated and the women. No prefacing. No apologies. This was just who she was. I described the women I had dated and admitted being attracted to men. I was empowered, bold and daring, safe. Once I admitted this, I experienced a gentle lightening, accompanied by the physical sensation of cool vapor with shades of cream and Alice blue rising from my skin.

The following morning, I decided to come out, gradually, to avoid more turmoil than I could process at any given time. First, by coming out as bisexual to a handful of acquaintances. From there, I settled on coming out most comfortably as *gay,* a term that still zinged as it passed my lips, but any discomfort was easily offset by the gentle golden buzz of the Y that lingered in my chest, the familiar and eclectic chartreuse of the G, and the emboldening effect of red bleeding into magenta of the A in between. Coming out became easier every time I told someone. When I visited Steve, Lee, and Alex, who were still entrenched as graduate students at the University of Florida, I met a friend of theirs who was charming and relatable. He was gay. I was gay. They invited me to join them for the university's homecoming celebration, so I used the opportunity to ask their friend to come as my date. To my relief, he said yes and I never transformed into an immoral creature filled with an abominable predatory lust. I remained human.

I had never really shared any information about my romantic life to my parents, let alone my siblings or anyone else in the extended family. It seemed completely irrelevant to have to come out to them at the time. There was no dissonance in this decision, no conflict with my own sense of authenticity. Though I mustered up what gumption I could to come out to Lee, Steve, and Alex, who were by now like brothers to me. Over the phone, I started with Alex, who was good friends with my date. He was excited. Next, I came out to Lee. Without skipping a beat, he reassured me in a natural earnestness that it did not make a difference, and we continued on with our conversation. Steve was perhaps a little taken aback because it put a kink in the image he had of what our group of four partly represented, a masculine brotherhood. But with some minor hesitation, he eventually came around.

Coming out to other medical students in my class was less intimidating than I feared. After I came out, other medical students, going through their own journeys, decided to come out. Our group grew from two out medical students, to three, then four, then five. Over the final months of my first year in medical school, I became closer with Johnny and Vigi, who were excited to introduce me to the invisible gay nation I had been missing. They took personal responsibility on behalf of the gay community to tutor me in my introduction to local gay culture, its lingua franca, its faux pas and social centers. With each experience, I gained a greater appreciation for a newfound level of comfort and freedom I found invigorating and enlivening. I was male. I was American. I was Hispanic. I was a synesthete. I came from a low-income family. I was gay. I spoke Spanish. I spoke English. I spoke science. I spoke pop culture. I spoke. I came to embrace the full anatomy of my identity and all its intersections.

In the fall of my second year of medical school, I acted as lead on a trip to Haiti. The focus of this particular trip was to support the Ministry of Health in providing annual medical screenings in Thomonde, a rural commune of central Haiti. Our group settled into an open-air roofed enclosure near a local school. On the morning of our second day there, I awoke with a headache. I had been prone to migraines as a child, but this was different. It was static and focal, a radiating, almost lancinating pain located over the right side of my scalp. When I bent forward to tie my shoes, the pain grew and spread. I stood up and, after a few minutes, the pain improved slightly. I ran my fingers along my scalp and felt a slight divot. I could not discern with confidence whether it had always been there or had slowly developed over time without me realizing it. Either way, the divot was prominent. The harder I pushed, the easier it was to find inconsistencies in the firmness of my scalp over it. The pain increased the closer I got to the pit.

My headache continued throughout the day. Though I knew this was abnormal, I was unsure whether I needed to consider it urgent.

Three days later, the headache persisted, so I went ahead and mentioned it to one of the attendings on the trip, Dr. Brad

Gerstner, a renowned neurosurgeon who had founded the organization that sponsored our medical trips. Unlike the other surgeons I had met, he seemed uncharacteristically charismatic and approachable. It also helped that he remembered me from a previous trip, where I won a spirited impromptu dance-off with a local village shaman to the blazing backdrop of rhythmic *rara* music. The shaman crowned me with a shared toast of homemade clairin, the strongest rum I have ever had. Dr. Gerstner started referring to me as "the second-best dancer in Haiti . . . after myself of course."

I approached Dr. Gerstner casually as if I were asking advice about a patient I had just seen. "Say, what does it usually mean when someone develops a new sudden headache almost out of nowhere?" He looked at me over the frames of his glasses, which were positioned halfway down the bridge of his nose. "It usually means they're going to die," he replied.

That was concerning. But I was still awake and alert. In fact, aside from the headache, I did not have any new symptoms. Just to be sure, after a minute of silence while we finished writing some notes, I told him I had developed a "new sudden headache almost out of nowhere."

He brought the tip of his Montblanc pen up to his lower lip. "It's probably nothing. Let's get it checked out when we're back in Miami."

Two weeks later, Dr. Gerstner invited me to his office. He introduced me to his two administrative assistants, his right and left hands, two bubbly middle-aged women. They placed a plastic patient ID bracelet on my wrist and escorted me through the hospital from one test to the next for a complete evaluation. Blood tests, X-rays, and an MRI with and without contrast.

A few days passed and Dr. Gerstner invited me back into his office. "We're actually not sure what it is," he said, pointedly. "It doesn't look like a dermoid cyst. But you have something at the edge of your skull and the right parietal portion of your brain. I'm going to refer you to one of the other neurosurgeons in our group who specializes in this sort of thing. He's early in his career but good."

I was unsure of what to conclude from the exchange but felt comforted that I would at least be able to continue the process of moving toward an answer.

Sitting in the clinic room of what had become my official neurosurgeon, I watched as he tugged back on his surgical cap and put his hospital-branded ballpoint pen back into his white coat pocket. "I'm not sure what it is and neither are the radiologists," he admitted. "At least we don't think it's vascular."

"Are you sure?" I asked trying not to overstep my bounds as a half-patient/half-medical student. "It seems somewhat gravity dependent, and I can feel the pulsation when it hurts the most."

The pain had been fluctuating from four out of ten to about a seven out of ten.

"Let's just do some watchful waiting for now. We'll do another MRI in a month and see if anything has changed."

A month later, the same clinic room. "It looks like it's grown. We're still not sure what it is. Is it still hurting?"

"Constantly. It's now mostly in the range of seven out of ten. I don't want to complain, but it's getting kind of challenging to focus in class."

"Hm. I think at this point it makes sense for us to do an excisional biopsy."

"What would that entail?"

"I would go in and use a small curette to scrape off a small sample to see what it is, and if everything looks okay, after we get a wet read on the frozen section from pathology, we'll go ahead and remove as much of it as we can."

"What are the risks if we don't?"

"It'll probably keep growing, but it's hard to say because we don't know what it is. It's even possible that what we think is growth is just a difference in how the MRI slices were obtained. We met on the tumor board to discuss your case, and one thing we're confident about is that it's definitely decalcifying the surrounding skull, which is why you have that divot in your scalp. Over time, it'll likely eat away at most of the surrounding skull."

He then went into the preoperative script of all the potential risks tied to any surgery, ranging from infection to death. The most alarming danger was that, if the lesion was vascular, there might be a chance that it would bleed, which could potentially be catastrophic.

I continued to study through my courses as usual. I wondered whether the surgery could have any potential effect on my synesthesia. I was just starting to get a better handle on understanding it. Could the mass somehow be tied to my experiences? I could not conceive what the world would be like if I lost it. It would be like excising a sensory organ.

I received a call mid-December that an operating room slot was available on the day after Christmas. I thought it a fair negotiation that if anything adverse happened with the surgery that I would have at least been able to spend one more Christmas with my family. The night before the surgery I stayed at my parents' house. My mother offered to shave my head with the same worn pair of clippers she had used to cut my hair as a child.

Lying on a stretcher in the early morning of the surgery, dressed in an oversized patient gown, I was not sure what would happen next. I told my parents that I would see them after surgery, but I knew that was not something I could promise. Still, I was determined. My scalp on the right side of my head was marked with a purple surgical pen. The anesthesiologists came up and introduced themselves and inserted a large-bore catheter into each arm in case they needed to administer blood rapidly. I eyed their technique for future reference. They gave me a small bolus of midazolam, a benzodiazepine known for its amnestic properties. Knowing that I would not remember much more after the infusion started, I reflected one last time on my life and gave a silent thank-you to the universe as I smiled and told my parents that I loved them.

I stopped existing, and then I existed once again.

First, in a universe filled only with the faint sound of two women's voices in distant conversation. The voices then came with a fluttering of pastel colors, interrupted occasionally by a robotic tangerine "boop." Then there was pain. I was wholly embodied by a white-hot pain that radiated at the top of my head as if someone had laid a tall stack of cinder blocks on me, then a crinkling black and red-orange seething pain in my urethra, the result of the urinary catheter. I reached down to stop whatever was causing it. One of the voices grew closer and more intelligible, "That's okay, honey! Don't touch it! Just go ahead! You have a catheter in." I dropped my arms to the side and opened my eyes. Light slowly bleared into focus. I realized I was in the post-anesthesia care unit. The operation was over. I had no idea what had happened or what to expect, but my first rational conscious thought directed me to look for writing to confirm my synesthesia was still with me. I looked up above and saw the placard that read:

POST-ANESTHESIA CARE UNIT

purplewhitecarrotmontanadustblackredtawnycobaltcarrot
montanadustautumnblazebluecarrotaliceredcrowredraven
cobaltlemontawnyalicemontanadust

All the colors of the letters were still there exactly the way they were before the surgery. My eyes drifted closed once again to rest.

I am unsure of how much time elapsed before I was in my hospital room, drowsy but awakening. My family arrived as did Lee, Steve, Alex, and everyone's respective girlfriends. There was more activity around me than the pain shooting out of my head allowed me to comprehend. The cinder blocks were still stacked high on top of my scalp. After what felt like hours, I was finally given pain medication, which dulled everything, including my surroundings. I was grateful for a brief reprieve. Later in the evening one of the surgical residents arrived to check in on me and give updates. The frozen section revealed a benign angiofibroma. *Benign* was all I needed to hear. The surgeon was able to excise the entirety of the mass and replace all of the missing skull with a combination of titanium mesh and polymethyl methacrylate, also known as "bone cement," an unfortunate misnomer. Over the right side of my scalp was an oblong inverse U-shaped incision stapled shut with a protruding drain dangling out of it. A plastic bottle at the end of the drain slowly filled with blood, like sap from a maple tree. They would keep me as an inpatient until the drain stopped leaking.

In the meantime, I continued my hospital immersion, this time from the perspective of a patient. If you are admitted to the hospital, you rarely see or have ready access to your providers unless your illness is "heading in the wrong direction" flagrantly toward death. You spend hours in bed with what look like plastic flotation devices strapped around your calves, inflating and deflating of their own volition. Typically once a day, in the very early hours of the morning, you see your physician for a few minutes as they run in and out to make sure you are alive and healthy enough to be discharged. When they finally pulled the drain out of my head, I was discharged with a plastic bag full of thin, beige, elastic hospital stockings knotted on one end to cover my incision site and prevent infection. Unfashionable, but practical.

I also had an opportunity to learn what had gone on in the operating room. In the weeks leading up to the surgery, another medical student whom I befriended mentioned that he was familiar with my neurosurgeon. He had been scrubbing into some of his cases in figuring out whether he would pursue a residency in neurosurgery. I told him the date and time of my case, and he was granted permission to scrub in. He told me how he arrived a few minutes late into the case, just in time to see my scalp opened wide around a pulsating mass that had replaced much of the surrounding skull. It was a vascular tumor. There was trepidation in the room to get near the mass. They ultimately called one of the most senior neurosurgeons in the hospital to evaluate the situation. He came in and scoffed, "It's probably some malformed bundle of blood vessels that was there from birth and just now grew enough to eat away at the skull. It should be fine to just scoop it out like ice cream."

Another medical student in my class happened to be working with a radiologist who discussed my case with a colleague. They, too, were perplexed by the available images of the mass. They suspected it was vascular, though they saw only subtle contrast enhancement on vascular imaging. They were, however, fairly convinced not only that the tumor was malignant but that it also extended toward the surface of my brain.

I was so grateful for the care that I received. But I was even more grateful for what I learned from my own medical case. I vowed never to forget the experience from the side of the clinician or the side of the patient as I made my way through medical training. The day that I returned to class, I still donned my beige knotted sock on my head. Eventually, the wound healed enough that I removed my head covering to reveal the incision. It became a prominent scar, a precious memento. I wanted to make sure that I would be reminded every time I looked in the mirror and every time I saw a picture of myself: we are all patients, we are all healers, one and the same. As I entered into my third year of medical school, I kept my head shaved when stepping out of the classroom and onto our hospital wards to care for patients—wounded, but ready.

Throughout my time on the wards of the hospital, I became well acquainted with what it means to be in a multicultural environment. The variety of cultural backgrounds blended from clinical clerkship to clinical clerkship. I made the active decision of attempting to structure the order of my clerkships to parallel the progression through the stages of life, starting with obstetrics and gynecology. I already had some basic exposure to labor and delivery, to say the least, though the raw diversity of language and cultural nuances around each birthing was endlessly unique. Pain is experienced by each culture so differently. One woman might sputter a staccato of choice curse words in English while another woman might braid long daisy chains of curse words in Spanish that rolled effortlessly off her tongue. The obstetrician's coxswain call was "Push, Mami! Push!" One woman hiccups "Aya!" followed by a hissing pulmonic "Iiiiiitthhh!" with each contraction. Another wails "Waiiiiii!" while yet another exclaims to every appearance of the Virgin Mary in a verbal dervish, "*Ay, Dios mio, Santa Madre de Dios, Jesucristo, Maria de Gaudalupe de la Caridad del Cobre de los Ángeles de la Imaculada Concepción!*"

The various patients I met during medical school, from beginning to end, would form the foundation for how I later interacted with patients as a physician. I was assigned a patient to follow by the two chief gynecology residents responsible for the whole ward.

The patient I was assigned, a woman in her early thirties, had been on the service (ward) for almost two months. She was the first patient I met with widely metastatic cancer. Her ovarian cancer had spread to the majority of her intestines and the folds within her abdominal cavity. To prolong her life, she agreed to neoadjuvant chemotherapy, opting to undergo rounds of chemotherapy before having her tumor removed in a radical procedure called pelvic exenteration, also known as a pelvic evisceration. During the procedure, surgeons removed her uterus, ovaries, rectum, bladder, and urethra—all of her pelvic organs. Nothing remained except for the remnants of a recurring metastasized malignancy that had somehow found shelter within her lymphatic system. Her bodily functions below her chest were replaced by plastic tubes and catheters, which emerged from a series of holes in her skin. Though the chemotherapy briefly staved off a recurrence of her cancer, it led mostly to an excruciating delayed neuropathic pain syndrome. Despite mounting doses of analgesics, she lay in bed with constant nausea and irreconcilable pain. Whenever I greeted her, I felt the weight of her despair and her heavy regrets. I was face-to-face with her agony.

I wanted to do something, anything, to help her. I researched all of her medications and her analgesics to find the ideal combination that could provide her some relief without sedating her to the point where she would have difficulty breathing on her own. She didn't want hospice care. Her mother and her husband refused to let her even consider it. With few options left, I decided that I could do one more thing, maybe not as a medical student, but as a human. I sat with her in the afternoons when I was not in the OR. I sat with her and just listened. I had nothing I could say to her, but I could sit, and nod, and bear witness to her existence, to let her know that I would remember that she was here. She spoke in a soft whimpering mix of English and Spanish that, in moments of particular stress, would often crack. During such moments, she would bring me in close and remind me that I knew *nothing*. That I had yet to grow and that one day I would find out that life was painful.

We would sit there in her desolation together. She would hold my hand and wonder if I understood. I wondered if anyone who had never experienced death firsthand could ever understand what it felt like to die. My experience with her would be the first of many that determined how I would connect with each patient. By nature, I felt like a machine, one with a reliable operating system that issues commands for my body to execute. No questions asked. A warm, pulsating, unfeeling circuit board quietly defragmenting until the next command was issued. Yet, with the synesthetic sensations bestowed by the mirror-touch synesthesia, it was as though the computer had also been programmed to feel "emotions" or at least the physical sensations that often represent emotions. It was up to me to figure out what to do next, how to interpret these felt emotions.

However, a remarkable change began to occur, one where the mirrored sensations re-created a mirrored experience, which became a hyperbolic replacement for empathy, which in turn became an unrelenting engine for profound compassion, a compulsory drive toward kindness. One morning our team was rounding near her room. As another medical student presented to the chief residents the updated clinical details of the patient they were following, I reviewed my notes. I was up next and I had observed that the less you read from your notes, the more likely the assumption that you "know your patient" well. I flipped pages filled with my scribbles on my clipboard trying to memorize every last lab value when in the corner of my eye I caught sight of my patient in her room. The door was cracked open and the curtain for her bed was pulled half open. She was lying flat on the bed. I felt the sensation of her suffering, which held slightly less tension in her facial expression. I mirrored the sensation of dry chapping lips on my own, my tongue reflected the slightly protruded sensation as it stuck and dragged along the desiccated skin of her lips. This suffering was also slightly different compared to what I had experienced with her before. I felt hopelessness as well as a gnawing physical thirst. There was a small clear disposable cup on her meal tray next to an unopened paper menu folded in half. The meal tray was several feet away from her bed. I watched her tilt her eyes toward the table. Before the thought entered my awareness, I had stepped into her room. Before the thought was completed, I was elevating the head of her bed. As she swallowed, I felt the sensation of a few drops of water slide down the sides of my mouth and neck. The brisker movements of her tongue and mouth were reflected in my own body as well. We simultaneously let out a quiet sigh of relief through our nostrils. She whispered in her usual labored voice, "Gracias . . . I was tired of waiting for help and gave up." I laid my hand on hers, looked in her eyes, and smiled. I gave her hand one last gentle pat, then walked out of the room and slipped back into rounds. It was my turn to present. At times, it was almost impossible for me to discern whether my acts of kindness were in response to the need of the patient or to cease the echoed pains and discomfort I felt within myself. Given the circumstances of my training, however, rather than writhe in this psychic anguish, I assumed that we were one and the same. That was acceptable enough.

During my internal medicine clerkship, I again had a short rotation on an inpatient oncology unit. Because it was a hospitalist service, I was able to work directly with the attending rather than with a resident. Dr. Cynthia Garcia, the attending I was paired with, had just finished her internal medicine residency as well as her year as a chief resident. She was bright, smart, relatable. Dispensing with the hollow honorifics, she insisted I call her "Cynthia." I was so appreciative of the care she took in making sure that I understood each patient, the fundamental concept behind the disease, and the treatment ordered for each disease.

She was patient enough to sit quietly as I tried to re-create for her my understanding of the pathophysiologic mechanisms at play at the molecular level from the inciting event all the way up to the clinical manifestation, which seemed to me the only way to truly make lasting changes to my framework of medical logic.

It was with Cynthia, on her last day, that I experienced the full force of my first death in an unexpected code blue. With each rotation thereafter, I had to focus on my training as a clinician and my training as a synesthete. I had to practice keeping everything in order. Until I could get a handle on this much, I couldn't start anticipating what to do as I resonated with the patient's experience.

I had to take in each experience to its fullest extent, down to the level of subtle nuance, and allow myself to exist as fully as I could in the perspective of the patient and the surrounding environment, well beyond the fringes of my "little self." Then, I needed to grasp the coalesced emotion that arose in me, tinker with my sensory and emotional experiences, examine the amalgam of both under different lenses and lights, and triage quickly whether I needed to act now, process further, or place it on a shelf to address later.

I trained so intently that by the time I arrived at my trauma surgery rotation several months into my third year I felt much more prepared. The rush of the experience was thrilling. I had read and reread each relevant surgical tip for the management of trauma patients. I tried to get in as close as I could to each patient, as close as possible to the floodlight of senses, challenging myself to lean in far beyond my own comfort. I was ready with supplies and instruments in hand, anticipating the next order called out by the surgeon running the trauma code. I worked on increasing the accuracy and precision of my predictions. I wanted to be poised with exactly what the patient needed before the thought solidified in the surgeon's mind.

In the trauma center's frenzied chaos, I learned the importance of shifting my focus the moment I recognized the patient's physical experience was starting to overwhelm me. Once, a patient was brought in with a gunshot wound. After passing his initial trauma survey, he began to wake up from his sedation and, as he got less drowsy, grow more agitated. The nurse caring for him asked for more sedatives and restraints, but the surgeon responsible for writing the order had just been called away for an emergent operative case. The nurse had a solution. She casually walked over and told the wounded man, "Here, hon. Let me help get you comfortable." She brought up the blankets to his chest, then proceeded to tuck him in as if he were going to bed. As she continued to tuck the sheet further and further under the thin mattress, I realized, she was hog-tying him down. I felt the weary dimmed lag of the nurse's facial muscles and the wary contracted elevation of her shoulders, which stored tension and exhaustion. Confused, the patient struggled against the sheets like a fly trapped in a web. I felt the tensile strength of the blankets gripping my torso. The mirrored sensations began to swell in me just as three new trauma cases rolled into the emergency room. To focus, I turned my gaze toward his intravenous saline drip, concentrating all my attention and energy to each drip. I felt the small crescent of water along the edges of my mouth, running up my jawline to my ears. At the same time, I felt contained within the small cylindrical vessel, settled finally by the IV's persistent drip, before jumping into the next case.

The pace of the trauma center was intense, to say the least. Much of the trauma was the result of negligence, ignorance, or chance—hardly ever malice. The accidental and the random included traumatic amputations. A homeless man was once brought in after he was found screaming alongside a railway. He had been sleeping near the tracks, and the passing train severed his right arm above the elbow. When I entered the room to help treat him, I realized I couldn't feel sensations in most of my left arm. Because I wasn't aware of the man's injury at the time, I couldn't figure out why my arm was numb. I started surveying the room. A set of bloody towels was draped over the man's right arm. I turned around and suddenly felt a cold stillness in my arm accompanied by a distinct sense of fraying up near my bicep. Lying on a surgical tray was the man's amputated arm. I tried to focus my attention elsewhere. But when I turned back toward the man, the doctors had removed the bloody towels from his arm, exposing his amputation. Immediately, I felt below my shoulder the wet sinewy strands and clumps of muscle shredded by the train. I looked toward the automated dispensing cabinet, a vending machine for surgical supplies. I let my eyes dart and search through its contents, the various edges, points, and labels of tubes, tubs, clamps, sutures—all of which were neatly stacked in rows and wrapped in sterile packaging. Through this illusion of order, my sensations, though still fragmented, started to soothe. I took a few deep breaths, returned my attention to the man's wound, anchored myself to the feeling of order, then catapulted back into the action.

At another moment, a teenage boy was brought in from a car accident. His older sister had lost control of the vehicle. Her friend in the front passenger seat died on the scene. Because the boy wasn't wearing his seatbelt, he flew through the windshield. His face was swollen with bruises and covered in lacerations. He was developing new bruises along his flanks. The senior trauma surgeon and the trauma fellow disagreed on how to treat the boy. The senior surgeon wanted to complete the trauma survey while the trauma fellow wanted to break protocol and take the boy immediately to the OR. The boy's blood pressure was dropping faster than he could receive blood. The decision was made to take him to the OR. I pushed on the right side of the stretcher while the senior surgeon pushed on the left. I glanced down and felt myself in the stretcher, my eyes swollen shut, bloodied, intubated. I had a similar feeling to the first death I witnessed, a hollow slipping sensation.

The boy was placed on the OR table. I felt the initial incision along his abdomen, which quickly faded into the hot and writhing experience of becoming his internal organs. I was a vessel, an open casket, filled with organisms soaked in blood. To find the source of bleeding, the surgeon performed an emergent exploratory laparotomy, but the hemorrhaging continued so fast that the boy went into cardiac arrest, slipping into a ventricular fibrillation. The trauma fellow applied metal paddles directly to his heart, quivering in his chest and in mine. His rhythm returned briefly, and then as soon as we both begun to breathe easier he went back into the fibrillation, failing to regain his rhythm a second time.

Days passed and I pressed on with the trauma surgeons. One night I treated a man hemorrhaging internally from a gunshot wound. The bullet had passed through his abdomen. To find the source of the hemorrhage, the surgeon on call cracked open the man's thoracic cavity. When the man went into cardiac arrest, the surgeon attempted a cross-clamping of the man's aorta. To continue providing blood to his brain during the process, the surgeon asked me to perform an internal cardiac massage. I placed my hands around the man's heart, literally beating it for him. As I pumped his heart with my open palms, I felt my hands in my own chest pumping my own heart, keeping myself alive. If I stopped, I remember wondering, would my heart stop, too? After about twenty minutes, the code was called and time of death was pronounced.

When my trauma clerkship was over, I wanted to continue in trauma surgery. I was enamored of the structured protocols and algorithms, the machinery of the process. I had grown confident enough that I could even handle the challenge of my synesthetic experiences. Though I knew full well that the exhilaration was yet another medical student trope, a "trauma surgery rush." You needed to give yourself a few weeks to come down off the high before deciding whether trauma surgery was meant for you. I continued in my clerkships, experiencing, learning, giving myself space to fully contemplate what a career in each specialty would be like. That is, until I arrived at the last clerkship of my third year, neurology.

Because I had a creeping suspicion that neurology was my destined home since my second year of medical school, I saved it for my last rotation when I put together my third-year schedule. One of the lectures given during the second-year neurosciences module was a nontraditional one by Dr. Steven Sevush. It was about the split-brain phenomenon and its implications in understanding consciousness. It was not a typical medical school lecture, but that's why it was so appealing. Of all the lectures I had in medical school, that was the one where I could feel my entire body come alive, buzzing and revving. I had goosebumps, chills, butterflies, all of it. Neurology ascended to the top of my list. But at the start of the third year, I felt I had to put it aside. I wanted to prevent my interest in neurology from interfering with my curiosity to learn other specialties.

I felt vindicated as soon as I walked onto the ward. The neurologists were geeky and awkward. They were endlessly curious, insatiable in their lust for learning. I took my time with my patients to get to know them and observe their behavior, patiently allowing them, their exam, their brain to converse and share with me what was going on below the surface.

One of the patients on the neurology ward was a middle-aged woman who had a small ischemic stroke in the left part of her brain. Albeit "small," the stroke's impact was enormous. She was able to retain most movement in her right arm and leg, but she had completely lost the ability to speak. She could follow most directions, but when asked questions or asked to repeat a phrase, her face contorted and reddened as she tried to speak. She produced a few sounds or chunks of words, but they were not always relevant to the question. After the first few attempts, she threw her hands into the air and shook her head in disapproval of her situation, her body, her life. Her hands landed in the bed. "Is there anything else we can do for you?" one of the neurology residents asked. She took in a breath. I felt the sensation of her face furrowing again as she tried to push her words through an invisible, impenetrable barrier. Her face relaxed. I felt a phantom tear roll down my right cheek as it rolled down her left. This was the routine each time we rounded in the morning and in the afternoon. No change. There were no speech therapists available at the time, and so she would soon be discharged to a rehabilitation facility as soon as a bed became available. "It'll be a few days before she can be transferred, so why don't you follow her?" offered one of the neurology residents. "We're not doing much for her now since her work-up is done. Maybe you can keep her busy while she's here."

Perhaps the intent was to keep me busy, but she was one of the first patients I was assigned to follow on the ward. I followed her in earnest. I observed her closely, patiently, deliberately. The more time I spent with her at her bedside, the more committed I was to help her get her words out, if only a simple phrase. She could comprehend. She could form some syllables and sounds. She had lost her speech, but she had not lost her motivation. Each time she tried to get a word out, her frustration sprang to the surface. She was angry. I did what a medical student does best and researched the causes of her deficit, her *non-fluent aphasia,* and sought out any skills—commands, prompts, maneuvers—that I could use to make progress of any kind. With each exam, I added a set of new techniques until after few days, *we*—she *and* I—were able to demonstrate how the woman who could not speak actually *could* speak, if you gave her the priming she needed with common sequences. Presenting our progress to the team, I started, "1 . . . 2 . . . 3 . . ." She angled forward and with minimal force brought out "4 . . . . . . 5 . . . 6 . . . 7, 8, 9, 10." She was beaming. "11! 12! 13! 14! 15! 16!" She turned to the team excitedly, as if she were getting into the best part of a story, "17 . . . 18 . . . 19 . . . 20!" The team applauded. She gripped the folded edge of her hospital blanket with joy. I placed my left hand on her shoulder and shook her hand with my right. I leaned in, "Alright . . . Ready to show them the days of the week?" She bashfully turned away to her left as she giggled.

I repeated this process with my other patients. I showed how the man who seemed confused and agitated was not afflicted by a presumptive toxic or metabolic insult. He was actually disoriented by his dense left hemineglect, which led to a somatoparaphrenia, prompting him to search for what he believed was the disembodied left arm of his deceased older brother lying in his bed. It was, in fact, his own arm. Sir William Osler would have been proud. Or at least that was what I told myself. All that genuinely mattered to me was that I was having fun and enjoying every moment of it.

With the added confidence at the end of my third year, I decided that this would be as good an opportunity as ever to explore whatever other interests I had and investigate them in detail. I had not done any research up until that point, but other medical students had already found their mentors or were publishing proficiently. I enjoyed research enough that I felt the need to explore what life might be like as a clinician-scientist. If I was going to structure my career path for a goal such as this, I would need to be sure that this would be my dedicated academic objective. Otherwise, I might not be able to set an appropriate trajectory for myself. I applied to take a funded year out of medical school through the Doris Duke Clinical Research Fellowship Program. Knowing that a research mentor could make or break an experience, I sought the best research mentor available in the neurosciences. I found this in Dr. Peg Nopoulos, the mentor of mentors for the research program. She specialized in neuroimaging research in neuropsychiatry. There was just one small catch. Her lab was located far from Miami, geographically and culturally, among the cornfields in the heart of Iowa.

CHAPTER 3

A Piñata Full of Tricks

I DROVE ACROSS THE IOWA STATE LINE AROUND SUNSET. THE landscape transformed into a warm gold, the vast stretch of wheat fields bathed in the setting sun. As the light continued to fade, fireflies appeared on the road, a bit of magic spreading and swirling gently in the light breeze.

Once I settled into my apartment, a spacious two-bedroom across the street from the University of Iowa's research hospital, I immediately started working with Peg Nopoulos. A disciple of Dr. Nancy Andreasen, a National Medal of Science–winning neuroscientist and neuropsychiatrist who developed some of the earliest brain-imaging studies of schizophrenia, Peg was a brilliant investigator, studying childhood neurologic and psychiatric diseases through modern neuroimaging techniques. She was so invested in seeing me develop as a clinician-scientist that she called herself my "research momma." Her wild red-headed curls with silver half-rimmed glasses framed the mix of scarlet, magenta, violet, and a light orange from her 2s, occasional 3s, and little 5.

Living up to her numbers, Peg was usually bursting with excitement. She could have just as easily blended into a local diner as a rambunctious waitress as she did as a leader in Iowa's neuroscientific community. I planted myself in a small office with only one directive: study, explore, question, and fail all I wanted.

Under Peg, I met regularly with researchers in the university's neuroscience department, who specialized in visuospatial ability, executive functions, and empathy. Not only was this a crucial step in my training as a neurologist, it just happened to have the added bonus of helping me make sense of how all three areas relate to my synesthesia. When I initially opened up to Peg about my trait, she offered her own hypothesis. She told me about evidence suggesting synesthetes like me exhibit a higher degree of hyperconnectivity than nonsynesthetes, specifically in the V4 region of the brain's occipital lobe and the adjacent fusiform face area, which helps synesthetes and nonsynesthetes alike recognize faces and graphemes, familiar sets of images tied to specific meaning, especially letters and numbers. This kind of hyperconnectivity, Peg explained to me, is most likely the result of defective pruning of different synaptic connections or a decreased inhibition of synapses within different regions of the brain.

This idea of defective pruning was particularly interesting to me because pruning is a natural part of the brain's development as it matures from childhood to adulthood. The connections between the brain's different regions are most intense in early childhood when we are trying to absorb as much new information as possible. Around puberty, however, in an effort to increase the efficiency with which information travels between its different regions, the brain starts to prune and refine these connections, decreasing useless or unnecessary "noise" that might slow down crucial signals traveling through the brain's labyrinthine wiring. Because synesthetes usually have two or more forms of the trait—as I would soon discover I did— hyperconnectivity also suggests that the biological mechanism responsible for synesthesia isn't concentrated to one specific location in the brain. Most likely it is instead distributed variably across the brain, like the spotted coat of a calico cat.

In its simplest and most fundamental form, the nervous system connects the external physical world to our internal world, our "little self." The mechanics and nuances of this elegant connection are as marvelous as they are bewildering—for synesthetes and nonsynesthetes alike. Following our instinct to understand who we are, we are driven to understand the connection between these two worlds, whether as scientists, philosophers, artists, farmers, athletes, or children gathered around a campfire asking big questions under the night sky. This curiosity is an essential part of being human; it is embedded deep in the story of our "selves."

How, though, do we begin to grasp the enormity and dimensionality of the living fabric between the external and internal worlds? To understand visuospatial ability, I relied on the *information-processing model.* Though imperfect, this model at least affords us a more concrete diagram of these abstract mechanisms. It is a deceptively simple three-step process that starts with an experience, or stimulus, followed by an evaluation of this stimulus, which then leads to an action, or response. The trinity of *stimulus-evaluation-response* is often translated into the language of systems as *input-process-output.*

In the first step, an input, or stimulus, is received and then perceived. Inputs can be any sensory information communicated through our physical bodies—receiving the world through light in the eyes, pressure on the skin, molecules on the tongue or in the nose, or change in the cellular ensemble of our tissues. To get a firm grip on the flow of this sensory information, I found it easiest to begin by taking the brain out of the picture and focusing on one of the neurologist's trademark tools, the reflex hammer. With the patient seated with their knees bent, feet flat on the ground, I tap with the reflex hammer just below the kneecap. I aim for the patellar ligament, and the tap causes it to stretch briefly and tug on the quadriceps. This triggers the stretch receptor end of a sensory neuron, also called muscle spindles, to send an electrochemical message up the sensory neuron into the spinal cord. Like falling dominoes, the signal triggers in the spinal cord the signal-receiving end of a motor neuron, which then relays the electrical signal back to the quadriceps muscle. This results in the familiar knee reflex kick. In addition to the synapse, the connecting site between the sensory neuron and motor neuron, an interneuron (triggered by the sensory neuron's signal) tells the hamstring to relax. This makes it easier for the quadriceps to make the leg kick. This entire reaction—tap (input), signal from sensory nerve to motor nerve (process), muscle contraction (output)—is called a *reflex arc*.

This all happens automatically. But what if we insert the brain into the process? While the principles remain the same, the entire sequence of events becomes substantially more complex. Though the input-process-output system remains, it is now linked together in the brain in four major areas: the thalamus, the cortex, the medial temporal lobes, and the basal ganglia. As the relay hub of information near the center of the brain, the thalamus connects the spinal cord and brainstem to the cortex, the outer layer of the brain that processes information. Curled under the brain at the intersection of the limbic and paralimbic systems, the medial temporal lobes make up different parts of the cortex that are most involved in memory and emotion. The basal ganglia are the brain's outgoing relay points between the thalamus and cortex that trigger either a motor action (movement or relaxation) or a nonmotor action (thought and emotion).

In tracing the signaling of input-process-output, our sensory systems receive information about our physical environment. This information enters the central nervous system through nerve-cell receptors embedded throughout our body. Because each cell is differentiated, each one is individually triggered by specific properties of the physical world. Thermoreceptors in the skin, for instance, react to temperature while stretch receptors in the stomach let us know when we're full. Baroreceptors in the carotid arteries react to blood pressure while chemoreceptors in the nose or on the tongue adjust to specific chemical or hormonal molecules. Each sensory receptor relays a specific signal through the cranial nerves or spinal nerves. Sensory information passes through the nervous system as an electrochemical signal and, for four of the five traditional senses, these signals travel through the thalamus before routing throughout the cortex. For the sense of smell, fragrant molecules land on olfactory chemoreceptors in the nose and signal directly through the olfactory cranial nerve into the cortex, bypassing the thalamus altogether.

Visual information is received in the occipital lobe (the back part of the brain), which integrates and relays this visual information throughout the cortex where it can link to different types of sensory information and, to give the visual experience context, different parts of the brain tied to memory and emotion. Learning about how the brain processes visual information, I was enthralled by the mechanisms involved in sight, how light first reflects off an object, then passes through the lens to the back of the eye, triggering the retina's diverse family of photoreceptors, which then bundles individual electrical signals in the optic nerves. These electrical signals travel back through the thalamus to the occipital lobe, where triggered patterns of individual neurons are relayed to the cortex. With its associative ability, the cortex then gives appropriate context to the signal.

Memorizing this neuroanatomy was one thing. Making this three-dimensional model of the brain useful meant understanding how it worked in real time. To grasp the brain's visuospatial ability in full, I regularly wandered through the corridors of the University of Iowa Hospital. On one such stroll I was focusing on how my brain generated footsteps when a door suddenly flung open. I reacted fast enough to step to the left, narrowly avoiding getting whacked in the nose. Later, I mentally debriefed how the visual information traveled through the brain's two major visual streams. The first stream is known as the ventral or "what" visual stream, which travels out toward the temporal lobes. As the light from the door entered my eyes and optic nerves, this sensory information was relayed through the occipital lobe, which then spread through the brain's neuronal networks tied to color and form, alerting me to a presence of a large brown rectangle. The association to form or shape likely occurred in the brain's fusiform face area, which helps us recognize faces and letters and numbers. From the brain's fusiform face area, visual information is relayed to the medial temporal lobe, which associates sensory signals to related memories, providing a rich three-dimensional and multisensory representation based on prior experience. This multisensory information is then relayed to language areas to provide an identifying "what" label for whatever it is we're looking at. Hence, my perception of "large mahogany door."

To perceive the location of the door in relation to my body, visual information traveled through the brain's second visual stream: the dorsal or "where" visual stream. Again, visual information about the door was relayed from the far back of the occipital lobe in its simplest form and, as the information was shuttled forward, it became associated with additional features such as form and motion, namely that a large rectangle was rapidly swinging toward me. In the parietal lobes, the top and back part of the brain, visual information about the door was tied to spatial representations in relation to other spatial representations. For example, the feeling of my feet on the ground or the sound of the doorknob turning to my right integrated to give me a sense of "where" the perceived door was in space as it rapidly approached my face. All these steps result in what we call sight—how we receive light from the world and then perceive it.

If this is how we *see,* then how do we *behold*? At the threshold of attention and awareness, we acknowledge, filter, contemplate, solve, decide, and react in a constantly expanding universe of permutations through a process of evaluation and interpretation. This process occurs partly through the association of visual information to representations of physical features such as color and form. As the visual information of a large mahogany door traveled through my memory and emotion-related areas of my cortex, my brain compared new information (an oncoming door) with previously stored information (all the other times I've run into a door or had a door open against me). This leads to association with expectations, emotional significance, and ties with specific emotions. My immediate perception of an opening door became a "DOOR!" coming toward me, followed by an anticipation of pain. Once imbued with rich context, the information integrates with an extensive system of neurons that are a part of the brain's executive control networks. These executive control networks are believed to be the primary mechanisms for evaluating sensory information and, through executive functions, generating a motor or nonmotor response. Which explains how visual information triggered the muscles in my body to avoid getting hit by the door.

Executive function refers to a set of brain activities made possible through the prefrontal cortex, the part of our brain behind our forehead, which works in tandem with interrelated regions across the brain. Like Samson's hair in the Old Testament story, our prefrontal cortex is a source of great power and great vulnerability. The most recently evolved part of our brains, the prefrontal cortex is also susceptible to impaired function and, as it develops from childhood to early adulthood, is one of the last parts of the brain to mature. The three core components of executive function are working memory, cognitive flexibility, and self-control. *Working memory* is literally the brain's ability to hold information, then manipulate or apply that information going forward. *Cognitive flexibility* is the neural basis for the brain's capacity to associate information with previously processed information such as associating that the heaviness of a cinderblock makes it possible to prop a door open in place of a doorstop. *Self-control* refers to the brain's ability to exercise voluntary and involuntary inhibitory control, which lets the brain select what we pay attention to while resisting or disregarding distractions, impulses, and temptations. This allows the brain to focus on goal-directed behavior.

Like pruning, it is also possible that an above-average executive function helps regulate the chatter of thoughts and emotions. Increased brain activity and a more sophisticated ability to make loose associations between different kinds of stimuli, coupled with the capacity to filter through this increased activity and loose associations, could allow a person to more successfully process this flood of external physical and internal mental information. This filter might allow synesthetes (or anyone in similar circumstances) to make rational sense of the information and pick out whatever bits they want or need at any given time. I wondered whether a filter like this could allow for heightened creativity or a higher propensity for innovation. Conversely, what would happen if or when the mechanism breaks down? Could it also allow for a benign lack of behavioral control just as it could permit a rip-roaring psychosis with hallucinations?

Assuming that I had some degree of hyperconnectivity and

some existing protective factors in place to manage synesthetic experiences, Peg speculated that my recent angiofibroma had led to the conditions for synesthesia or an even more heightened form of synesthesia. Mechanistically, it was possible that a developmental effect led to a more active right cerebral hemisphere. Or, even more specifically, a more developed right parietal lobe—which is responsible for cortical functions at the confluence of our visual and somatosensory systems—fostered my synesthesia. While I wanted to agree with Peg, I couldn't help considering if the opposite was true. What if my angiofibroma delayed the development of my right parietal lobe, which impaired my ability to filter out associations between different sets of sensory information?

The brain's executive function allows us to captain our body and our mind and, by extension, our life. These responses loop through three major nonmotor parts of the prefrontal cortex: the anterior cingulate cortex, the orbitofrontal cortex, and the dorsolateral prefrontal cortex. These three circuits work in parallel to do the following: process and express emotions, regulate mood, and process pain (the anterior cingulate cortex); regulate appropriate and empathetic behavior in the context of changing emotional states and physical factors (the orbitofrontal cortex); maintain, organize, and manipulate information through cognitive control in the pursuit and regulation of goal-oriented behavior (the dorsolateral prefrontal cortex). Injuries or malfunctioning of these circuits are often what lead to neurobehavioral and neuropsychiatric disorders, including alterations in some of our most complex human experiences such as empathy.

Because of my own mirror-touch experiences, I was particularly curious about the circuitry of empathy, literally, how does it work? Even if I didn't have a name for my experiences, let alone an appreciation that they were any different from what others were experiencing, I found myself gravitating toward the study of empathy. Though I was surprised to discover how much of a challenge it was to even define *empathy*. I find the most accessible definition comes from Brené Brown, a research professor in social work and specialist in qualitative research methods. She describes empathy as "feeling *with* people," the ability to step into whatever mental space someone occupies to let them know that they're not alone, that you've been there before—in short, that you feel *with* them.

In the quantitative study of empathy, this complex human behavior is generally thought of as a *capacity*—that is, the *potential*—to *understand* or *feel* what another person is experiencing from your own point of view. The more vividly you imagine understanding or feeling the mental state of others, regardless of how accurate you are, the more likely you are to experience empathy as an emotion-reading reflex.

Compared to the light tap of a reflex hammer, the sensory information that informs empathy is highly variable and multisensory. These signals include visual information such as emotional tears, eye gaze, facial expression or posture, and auditory information, including the tone of the other's voice. After arriving at the brain's primary sensory cortex, multiple networks there process sensory information to add context and emotional significance. This includes contributions from the mirror neuron system.

By the time I came across the mirror neuron system theory, it was just starting to gain credence in neuroscience, despite initial skepticism. Today, in the wake of a growing body of evidence, the mirror neuron system is a generally accepted theory about how the brain works. As a whole, the mirror neuron system describes neurons that create a three-dimensional multisensory simulation of what people observe in other people. Mirror neurons were first described by Giacomo Rizzolatti and his colleagues at the

University of Parma who were studying the hand and mouth actions of macaque monkeys. Using visuospatial maps, Rizzolatti and his team measured neuronal activity in areas of the brain involved in motor planning and the integration of body sensations. By the genius of curious observation, they noticed that the same set of neurons that were active when the monkey grabbed a piece of food were also active when the monkey simply observed one of the investigators grab a piece of food. Like the macaque, people exhibit similar mirroring properties while observing physical actions, emotions, disgust, and even pain. The hypothesis, then, is that the mirror neuron system automatically processes sensory information by simulating a reflection of the other's physical experience, including facial expressions and body language. The resulting response to the internal three-dimensional simulation is thus our own emotional response, based on our own capacity to compare the experience of another person with an experience that we have had in the past, which is stored and available for recognition and recall in our memory system. In other words, the mirror neuron system is the potential mechanism for why someone watching a person fall off a bike will flinch or cringe as if it were actually happening to them. The mirror neuron system can also "leak" in the form of unintentionally mirroring the posture and facial expressions of another person such as crossing your legs in the same direction as someone you're talking to or picking up a glass of water at the dinner table at the same time as the person sitting across from you, as if you were staring into a mirror. This might also be the mechanism behind emotional contagion.

Rather than shedding light on my mirror-touch experiences, I mistakenly assumed normal mirror neuron system activity was responsible for them and, more egregiously, dismissed my experiences as even more unremarkable, more in line with everyone else's.

And yet, I still had lingering questions. Do we have to cultivate the empathetic capacity by deliberately creating a stronger understanding or feeling of the mental state of another? Or, is it a matter of cultivating the ability to *allow* ourselves to experience more empathy? What if we do both? How could we make the way we imagine the mental state of others more vivid? How could we perceive that we have a lot more in common? Would we automatically be more prone to experience empathy? What would happen at the population level if we each felt more closely related to the "other?" What would be the downstream effects? Furthermore, if we have a mirror neuron system, how do we know that we're experiencing our own emotions and not the perceived emotions of the other? What circumstances affect the integrity of the boundary between the self and the other? Can we reinforce that boundary? Can we erase it?

I came to the conclusion that the scaffolding of empathy, its anatomical and conceptual structures, only explains a portion of how we can feel for others. For example, yawning can spread through a group of people. A contagious yawn likely spreads using the same scaffolding of the mirror neuron system. This scaffolding is likely responsible, in part, for other rich emotional and physical responses, like brief vocal bursts of "Hm" while catching up over coffee or sharing emotional tears of joy while watching a mother and father holding their firstborn child. For the most powerful and immediate forms of empathy, it seemed as if a higher awareness of the other person's mental states were necessary. The feeling part of empathy seems to occur at the threshold between recognition of another person's internal world and sharing in that world, that is, simultaneously understanding *and* feeling the experience of the other. In a sort of symbiotic resonance, our linked experience can then flow through with intense personal meaning that can feel transcendent, a gift from the divine. And just beyond empathy lies compassion, the drive for action. To then perform an act with the intention of alleviating the suffering of another is *kindness*. Empathy requires feeling *or* understanding. Heightened empathy requires feeling *and* understanding. Compassion requires motivation to relieve suffering. Kindness requires action.

Rather than continuing to try to figure through my lingering questions about empathy and the mechanisms underlying my synesthetic experiences on my own, I admitted to Peg that I still had many more questions about my trait and the brain than I had answers. She encouraged me to follow my curiosity and research synesthesia more deeply in the hope that I might learn something about myself and perhaps even about the inner workings of the brain during my year studying with her. To begin chipping away at all that I had yet to uncover about my synesthesia, I met with reputable neuroscientists across the country, all of whom demonstrated a penchant for exploring abnormal phenomenological experiences and cognitive functions.

One of the first such experts I met with was Dr. Darold Treffert, a psychiatrist who had developed an international renown in *savantism*. Savant syndrome (or savantism) occurs when a person with what might be called a neurodevelopmental disorder also has an isolated set of prodigious capacities or abilities such as playing music or performing massive arithmetic computations at close to the speed of a computer. Treffert worked with the likes of Kim Peek, the savant who inspired *Rain Man,* and Daniel Tammet, a savant known as "Brain Man" for his exceptional ability for memorization and language acquisition, who happens to have a mix of autism spectrum disorder, epilepsy, and synesthesia.

Much like the interneuron in the reflex arc of the knee, the brain has neurons that, instead of activating, inhibit other neurons. It's possible that people with savant syndrome have impaired neuronal systems or whole regions of the brain that naturally inhibit the activity of other parts of the brain, which would then make their synaptic targets more active, leading to a prodigious level of cognitive capacity.

Treffert was the first person to encourage me to write about my personal experiences. I approached the idea with significant hesitation, unsure of how it would be received by others, how it might impact my reputation as a clinician-scientist, or how patients would perceive me. I convinced myself that communicating such an experiential phenomenon in words would seem so subjective that it could not possibly be of interest to anyone but me. Moreover, I had just come out about my sexuality. To come out once again so publicly conjured up far more risks in my mind than possibilities. I had yet to meet another synesthete, so I assumed that at best I would be contributing to obscure pop science errata. I would be better off shelving the idea for now, I convinced myself, and continue on an academic career in medicine, keeping to the straight and narrow, the predictable and anonymous.

But, to further my own understanding of synesthesia, I visited the University of California, San Diego's Perception and Cognition Lab, where I met neurologist Dr. V. S. Ramachandran and his talented graduate student at the time, David Brang. Both men showed a keen interest in synesthesia and other atypical sensory experiences such as phantom limb syndrome, thalamic pain syndrome, and body identity integrity disorder (sometimes known as apotemnophilia). Within a day of my arrival, David began taking me through a battery of psychometric tests to objectively quantify as many facets of my synesthetic experience as possible in a controlled lab setting. Most of the tests were structured, but some were prompted by Ramachandran's eternally playful scientific curiosity—just to "have a look."

In the basement of Ramachandran's lab, I was assigned my first task: sit in front of a computer while wearing headphones and watch a black screen. Every so often, the screen flashed a red X or a white square. Each flash would randomly appear with or without a simultaneous high-pitched beep. When I saw the red X, or heard the beep, I had to press the keyboard's spacebar as quickly as I could. Ramachandran and Brang wanted to measure my reaction time, speculating that because synesthetes had greater connectivity of signals between different sensory areas in the brain, their brains would be faster at recognizing when both the red X *and* the beep appeared and thus have faster reaction times for this combined audio-visual prompt than nonsynesthetes.

When I finished the assignment, however, they told me they couldn't use the data. Turns out, I was an outlier. My reaction time was not only faster than nonsynesthetes, but also easily at least three times faster than the average synesthete. Based off the synesthetes they had tested, did this mean that I had more hyperconnectivity or integration than the average synesthete? Or, was my reaction time faster simply because my muscle reflexes were faster? We simultaneously raised eyebrows, unsure of what the results meant. Answering these kinds of questions would require several follow-up studies with comparison to an even larger group of synesthetes. My inclination at the time was to be more willing to accept new information that fed into my belief that I was ordinary, that I was just like everybody else. I told myself a story that the results were likely just random variability or perhaps technical error.

On another task, David asked me to stand with my back turned to a blank white board. David then used a black dry-erase marker to draw letters of the alphabet on the board. When he finished, he told me to turn around when he gave the signal and—not knowing which letters or how many different letters were written—figure out which letter was most often repeated. His hypothesis was that a synesthete would be able to use their synesthetic color as a guide and therefore recognize the most populated letter faster than a nonsynesthete. If his hypothesis was correct, I would be more likely to see right away the letter with the highest concentration on the board, guided by color alone.

"Ready?"

"Sure."

He raised his stopwatch, "Go!"

I turned and immediately shouted, "Blue! E!"

David barely had time to stop his timer. On the board, there was a combined total of a hundred Es and As. There were only a few more Es compared to As. We repeated the task several times to ensure I wasn't randomly guessing.

Next, David brought up an eye chart on a computer screen. He had me stand a few feet away and read each line until I reached the line with the smallest print, just as you would at the optometrist. When I finished, David told me to take a step back and read the chart again. This continued until I ended up more than twenty feet away, far into the hallway, finally out of space. As far back as I was, however, I could always make out the smallest letters, even if they were slowly coming in and out of focus. As long as there was some semblance to the letter, it would still have its respective synesthetic color: chartreuse smudge for G; brown splotch, D; red blur, A. David looked piqued and said, "I think we won't be able to count this task since, again, you're a bit of an extreme outlier. We haven't had anyone step that far back for this one."

Another day we sat down in Ramachandran's office, which was decorated with all sorts of artifacts and equipment rigged for psychometric testing. It was as if I had walked into the neuroscience wing of a museum of natural history. Sitting in his office, we tried some additional tasks, including describing the colors of overlapping letters and numbers. In the example of a 2 intersecting with a W, along the left corner of the W, the 2 was still red and the W was still green. But I was dumbfounded at the point where the 2 overlapped with the W. Focusing in on that spot was slightly disorienting, like staring into an abyss right at the center of the overlap, which was both colorful and colorless. It was only red and only green and neither simultaneously. If I stared long enough, I felt I could lose myself in the void. I wondered whether this was due to the phenomenon of inhibition, where the brain tries to suppress information in order to make sense of it, a reflexive attempt to erase information that doesn't fit its predicted, wired model. Suppressing an inconvenient piece of information is at the core of most optical illusions.

On another task, David drew three equally spaced 2s on a white sheet of paper, lengthwise. He held the page in front of me and asked, "What color is the 2?"

"Red . . . Well, it's actually a little more complicated than just re . . ."

"Good. Now keep telling me what color it is, over and over again. Don't stop until I tell you."

He began to rotate the page clockwise from my point of view.

"Red . . . Red . . . Red . . . Red . . . Red . . . Hm. Brown."

From my perspective, the right side of the page went from the three o'clock position to about four o'clock when the color of the 2 suddenly changed from its usual violet-licked red to a tawny brown. I made a realization two seconds after the color had changed.

"Oh. It's an N now."

To my brain, the number 2, when rotated 90 degrees clockwise, looks like an upright N. I was impressed by how dramatic a change there was in the color based on the orientation of the grapheme as well as how subconscious, how involuntary, the change of color was. I speculated that my brain's fusiform face area processed the grapheme as that area had been wired to do.

Patterns of light entering my eye that could represent a 2 are red. Patterns of light entering my eye that could represent N are brown. My conscious awareness of what I was looking at was less of a factor than what kind of light filtered through my optic nerves. My mind went through a synesthetic processing first, followed by a conscious awareness of what I was looking at, the latter part of my "what" visual stream. This was substantial, a phenomenon that I could really sink my teeth into because it felt less subjective and much more mechanical, more *neurological* than psychological, which was refreshing, given how the distinction is often blurred and illusory.

Curious about the connections between my optic tracts and the visual systems of my brain, Ramachandran asked me to stare at a black 3 on a white piece of paper for thirty seconds then look over at a blank white wall. When a person keeps his or her gaze fixed on a high contrast image, a residual photonegative afterimage lingers when the person looks away. This is one of the reasons why you sometimes get blinded by a camera flash or the sun. When I turned away from the piece of paper, the photonegative 3 persisted crisply on the wall in front of me. Ramachandran and David timed how long it took for the image to vanish completely. Even after I blinked, the afterimage lasted upwards of about three to five minutes. The average time recorded for nonsynesthetes, they told me, was fifteen to thirty seconds. Ramachandran speculated that a residual of images could persist through hyperconnectivity, making it easier for a neural signal to essentially "echo" in a feedback and feed-forward mechanism. He postulated that this was one of the additional mechanisms by which some people have near-photographic, or eidetic, memory. Maybe a person with such a memory can maintain a visual representation like photographic film with minimal to no effort through the brain's visual system, which automatically allows them to hold on to a visual image longer via this bidirectional echo instead of having to undergo the more involved task of deliberately encoding, storing, then retrieving the information in the memory systems of the brain.

Throughout my time in San Diego, Ramachandran, David, and I walked around campus in what he described as "sandwalks," an homage to Charles Darwin's idea-generating walks with his students. They asked more questions, each more tangential than the last. Nearing the end of a walk they asked if there was anything else that I thought was unusual about my sensory experience. By then I had already learned to question whatever I thought was normal or unusual, but I did have a sense that I had a tendency to be able to spot objects, particularly letters and numbers, more readily than others. We found a blank sheet of paper and I asked David to scribble all sorts of lines and curves haphazardly while I looked away. Without giving it much thought, I looked at the page and pointed out all the letters and numbers I saw, one after the other without stopping, even though David only drew a tangle of lines and curves. "A, F, B, C, Z, N, H, G, F, L." They were all laid out like colored pick-up sticks. Spotting them was simply a matter of attending to each color and recognizing its corresponding letter. I felt a little sheepish doing this because it felt like some kind of cheap parlor trick, which was why I was surprised by their level of interest in the phenomenon. David asked, "Were you ever good at the Where's Waldo or I Spy books growing up?"

"Those were some of my favorite books growing up, though I don't know if I was experimentally any better at it than anyone else."

We walked directly into a nearby bookstore and David purchased a stack of books, making sure that I hadn't previously read or looked through any of them. Back in the lab, he pulled out a stopwatch and prompted me to find Waldo as quickly as I could. "Go!" I tore through each of the Waldo books without stopping.

Waldo was practically jumping out of the page. "There." He turned each page. "Here. There. There."

We moved on to the I Spy. It proved to be a bit harder to identify a new object than spotting Waldo's familiar face. "Find a rabbit." But what did the rabbit look like? What was its color? Was it a wooden rabbit? A bunny? Or was I looking for an image of the March Hare at a tea party with Alice? There were limitations to the impromptu experiment. But it did provoke a few more questions from David and Ramachandran about how my visual system worked. Finally, David asked, "Do you happen to have mirror-touch synesthesia?"

"I've never heard of the term."

"It's when you feel the sensations of others that you observe as if it was happening on your own body."

I paused before answering, "Well . . . yes, but I'm pretty sure that everyone has that. Isn't that just what the mirror-neuron system that's been proposed is all about? I don't see what would make it so special as to merit its own name."

"Here, look at my face." He passed his fingertip along the left side of his face. "What do you feel?"

"A fingertip passing along the right side of my face."

"Yeah, that's not normal."

I slapped my forehead.

"Yeah, most of us *normal* humans do *not* have that happen to us," he confirmed, amusing himself. "You're just a piñata full of tricks, aren't you?"

David was kind enough to inform me about my trait, preliminary information I have since augmented with further research and experimentation. Mirror-touch synesthesia, experiencing the physical sensation of touch while observing touch, shares some of the more general properties of other forms of synesthesia. It occurs at the conscious level, meaning that a synesthete is aware of the process taking place and can even describe the details of the experience. Synesthetic experiences are triggered automatically by a visual stimulus not typically associated with the conscious experience: graphemes with color, sounds with tastes, and, in the case of mirror-touch synesthesia, sight with touch. For instance, I feel mirrored touch on parts of my body that correspond visually to whom or what I'm looking at, like in a mirror, especially when we are face-to-face. However, like in some mirror-touch synesthetes, the location can also be anatomical—left-to-left, right-to-right—when I am side-by-side with a person. I also experience mirrored touch more vividly depending on a few factors. My experience is heightened the more similar whatever it is I'm observing is to me. My mirror-touch sensations are closest to actual physical touch when I observe someone who shares similar characteristics with me, which is followed in slightly diminished intensity by observing someone who does not look like me. I experience mirror-touch sensations at decreasing levels of vividness when triggered by objects that appear human, like a mannequin, followed in decreasing intensity by objects that look humanoid, like an electrical outlet, which is tied to the common experience of pareidolia. Objects that have no clear resemblance to a person at all, like a glass of water, are the least intense, though I still experience them occasionally. Meanwhile, new experiences and unexpected experiences are heightened, like sudden surprise or shock or, more simply, when I'm caught off guard. Increased emotional significance, or salience, has a similar effect, such as when I see something I have strong emotional ties to, an experience not unlike I imagine victims of trauma go through. Attention also affects vividness. The more I attend to the mirror-touch experience, the more real it is. Conversely, the less I attend to it, the less vivid it is, though I am never really able to extinguish it completely. In this same vein, if I were to drink caffeine (greater activation) or drink alcohol or become sleep deprived (less inhibition) the experiences can become even more vivid because of the increased activity in the brain. These are all features that point to what imaging and EEG studies suggest, mainly that mirror touch is predominantly related to an increase in activation of the mirror neuron system, a near-constant trait linked to hyperconnectivity via cross-activation with the regions of our brain that ordinarily only activate when a person feels someone touch him or her. Additional support for this hypothesis comes from the observation of greater measures of structural brain connectivity in parts of the brain tied to social perception and empathy. Furthermore, there are fewer structural connections in parts of the brain involved in the perceived boundary between the self and the other, the most relevant area being the right temporoparietal junction, which sits at the intersection between regions that integrate vision, spatial perception, our body's mapped representation, memory, and emotional processing. Interestingly, this part of the brain is close to the location of my tumor. It's likely that the unique brain circuits of mirror-touch synesthetes are what cause them to perceive the space of their own body as more expansive, blending into the perceived body space of others.

As I learned about mirror-touch synesthesia, I kept returning to a few lingering questions: do mirror-touch synesthetes have a heightened or different form of empathy? Do *I*? Michael Banissy's group in the department of psychology at Goldsmiths, University of London, thoroughly validates each mirror-touch synesthete enrolled into their experiments with objective tests of authenticity, a commendable feat of methodologic rigor that reflects the importance of acknowledging wide variation in the synesthetic phenotype, specifically the expression of these traits. Validating cases of synesthesia is already a massive technical challenge to quantify objectively, much like quantifying the phenomenon of "empathy."

In his group's verified sample of mirror-touch synesthetes, Banissy found evidence that mirror-touch synesthetes demonstrate higher levels of *emotional empathy* than nonsynesthetes, though *cognitive empathy* levels were the same. This made intuitive sense to me, because the mirror-touch experience is largely embodied, and the trait's physical component might be more likely to evoke a strong emotional response to the experiences of other people. But what did this mean for me? I was having doubts about the practical significance of mirror touch. Was this just a meaningless quirk? Or, perhaps a vestigial part of my brain's wiring, like an appendix or an awkward tail? Whether there was a practical, measurable significance to my mirror-touch experiences would probably need to be determined in an experimental setting to ever be able to speak with truly objective confidence.

But then I reflected on my own experiences. Speaking from my own perspective, not as a scientist but as a perfectly imperfect and biased and uncertain human, I felt as though my mirror-touch experiences informed and bolstered what cognitive empathy I had, though it's challenging to imagine what my cognitive empathy would be like without my mirror-touch synesthesia because it's always been a natural part of how I sense the external world. To be able to *understand* another's perspective with 100 percent accuracy would border on telepathy. Mirror touch may not fit the standard form of telepathy found in the canon of science-fiction—but it helps to shine a light on those crucial moments in the hospital when asking *one more* question or asking the *right* question can change a life. Even if it ultimately only benefits a single person in my lifetime, that alone will have made my mirror touch completely worthwhile.

And what about the experience of *pain*? As a mirror-touch synesthete, was I truly feeling the pain of others, an echo of their pain, or was it simply my brain's own extrapolation of what their pain must be like? Given the phantom sensations I experienced, mirror touch felt distant yet similar to what I knew from the pain of a stubbed toe or a sliced finger. Self-perception of pain is a complicated subject. Any attempt to distill its relevant concepts into an easily digestible format is no less than a Herculean feat because there will always be a need to sacrifice nuances in exchange for pragmatic clarity. The more we engage in the art of distilling, the more space we generously offer up for interpretation.

The questions that arise from this particular question of perceiving pain are somewhat personal because I had to work through my own logical dilemmas and internal tensions with the subject. Initially, the research scientist in me balked at the use of phrases implying absolutes related to mirror-touch synesthesia such as "feels the exact pain of others" in an attempt to explain the phenomenon of mirror touch. After much deliberation, though, I eventually came to terms with this language.

That said, pain is incredibly subjective. In its scientific study, pain, like empathy, is typically measured via self-report. That means it's measured by the most elemental tool of investigation: asking. There are several validated scales and surveys with questions like, "Do you feel pain?" or "On a scale from zero to ten, how much pain are you in right now?" This is because there are yet to be reliable or valid tests for pain like there are for blood pressure and cholesterol levels. Even when we try to find objective measures in advanced imaging techniques that allow us to see small areas of activity in the brain via fMRIs, what we see is activation of somatosensory brain areas responsible for receiving and relaying information about sensations felt on the body and activation of regions of the brain involved in emotional salience such as the intensity of emotional or psychological experiences, including pain. What's even more fascinating is that even in people with a genetic insensitivity to physical pain, the brain activation patterns remain similar. In one study of brain activity, persons who watched their spouses receive painful electrical shocks showed activation in their entire "pain matrix" as if they were experiencing pain themselves. When these same people observed someone who was not their spouse receive painful shocks, brain activity patterns were similar, though not through the entire pain matrix. This possibly means the most we can interpret objectively—from either injury to our body, psychological distress, or observing someone else who is suffering—remains the after-effects of pain, which are ultimately psychological or at least predominantly reside in the domain of higher cognitive functions.

As an input, pain signals travel from the body to the spinal cord and then to the brain, leading to the interpretation of physical pain. Prior to the initial pain signal, cell damage from tissue injury causes the release of cytokines and other neuropeptides, which are a part of the inflammatory response, such as the cryptically named Substance P. Once released, these molecules trigger pain fibers responsible for conducting (from body to brain) the sensory experiences that we interpret as physical pain—be they thermal, chemical, or mechanical. These signals can then be processed in our brain for information related to location of injury, emotional salience, aversion, and motivation, which is arguably at the core of pain: "OW!" followed by the primal drive to stop the experience of pain.

Pain *feels* real and obvious. But pain is an illusion, unfairly so, considering the suffering it causes. The pain you feel in your index finger after a papercut is actually not in your finger at all. The pain is in the brain. Like other aspects of the input-process-output model, the brain creates pieces of a three-dimensional model that it ties to the nerve signals from the pain receptors in the finger, which then trigger a cascade of signals in the brain to conjure the illusion of pain in your index finger.

Quietly yet begrudgingly, I realized that I was looking at the experience of pain from an almost strictly medical point of view. Anything short of an exact replication of the distinct bottom-up thermal, chemical, or mechanical signals that I had intimate experience with—burning, stinging, aching—was not consistent with the word or the experience of "pain." However, after much reflection, it was apparent that I was using a very narrow and highly specific frame of reference. Likely realizing this too, members of the International Association for the Study of Pain (IASP) came to a similar conclusion in their consensus definition: an unpleasant sensory and emotional experience associated with actual or potential tissue damage, or described in terms of such damage.

Notice that a significant distinguishing feature driving the classification of "pain" is the experience of "unpleasant," also fairly subjective. That said, speaking from my own subjective experience, I do have a relatively higher threshold for pain to begin with, which may have an influence on my own perception of the emotional salience behind experienced sensations. However, the sensorial phenomena that I've experienced in relation to mirror-touch synesthesia behave within a fairly consistent set of overarching properties. In hearing the stories of other mirror-touch synesthetes, I feel more confident that most of these concepts still fit within generally accepted principles that describe the brain's functioning and the reception, perception, and interpretation of stimuli. The mirror-touch experiences seem more consistent with a top-down process—derived in the brain but experienced in the body—yet the experience of unpleasantness and, therefore pain, remain. Thus, people with mirror-touch experiences can experience other people's pain, pain derived from the pain of others.

The researcher in me would like to speak up, however, and qualify that, probabilistically, no two bodies are exactly alike, even in the case of "identical" twins. Therefore, the interpretation of pain is unlikely to be identical with absolute certainty from one person to the next, no matter how similar they may seem to be. While we cannot experience the same pain of another simultaneously, we can still experience a reflection or echo of that experience—a simulacrum of being "the other." It is entirely up to the beholder whether "same" and "exactly" are interpreted as a one-to-one correspondence down to the ill-defined subatomic level versus "close enough." As a scientist, I'm prone to endorse the former. Meanwhile, the rest of me knows that "close enough" can do just as well in most situations. Personally, the light tickling feeling from mirror touch all the way to more intense mirror-touch sensations can be unpleasant at times. There are situations where I have been caught off guard by the intensity of the sensation. You would think that a cell phone or pager was being mashed up against my face when it vibrated violently. When this does happen, I am yet again amazed at how easily influenced we can be by the functions of our brain. It is humbling. At the same time, each mirror-touch sensation is not necessarily interpreted into emotional empathy. I consider the sensations that do not lead to the full experience of empathy as components of what is described as somatic empathy (a sort of "mechanical empathy"). Even if the mirror-touch sensations can be somewhat mechanical at times, they always create an opportunity for a deeply meaningful experience of empathy. It's likely that the ability to attend to these opportunities for empathy can either be cultivated or ignored—voluntarily and involuntarily.

I have since broadened my original definition of "pain" to include its most essential elements. One of the aspects I love most about subjects like synesthesia and pain is how they emphasize the importance of a collaborative approach between scientific and nonscientific perspectives, balancing the relative internal-external validity of the subjective experience of being human and the objective measures that help us make sense of our individual worlds.

Later, during my stay in San Diego, as David and I continued to explore my synesthesia, he invited me to the university's magnetic encephalogram, or MEG, center. He wanted to get a better understanding of my brain's activity. MEG imaging takes advantage of electrical currents passing through the brain in a single direction, which generate a magnetic field with predictable properties. MEG measures the magnetic field generated, thus vicariously measuring the electrical activity of the brain and detailing where in the brain the activity is taking place.

I sat upright in a plastic throne-like contraption in the center of the room with my head inserted into a narrow and elongated tunnel. Over the course of a few tests, I looked at a series of numbers and letters with colors that either corresponded to my synesthetic colors or numbers and letters with colors that did not correspond to my synesthetic colors. The graphemes were each in front of gray backgrounds with varying shades. The goal of the experiment was to see whether the color vision and face recognition areas of my brain activated differently depending on how clearly the graphemes stood out compared to the background. This information could then help objectively define exactly where and when the brain crosses the threshold into synesthetic activity.

After sitting locked up in the magnetically shielded box, I stepped out for a break. David showed me some of the preliminary data. "It looks like there really is a bit of activity going on between the face and color recognition parts of your brain at the same time you see the numbers. As expected the activity is greatest when you see the numbers in your synesthetic colors. Interestingly, the sheer amount of activity taking place in your occipital lobe is far greater than any of the other synesthetes we've put in the MEG. But this is preliminary data, so I won't trust this completely until I actually analyze it closer. Again, you're a bit of an outlier here, too."

It made intuitive sense that visual information relayed through the occipital lobe before proceeding into the "what" visual stream would somehow "harmonize" with the hyperconnected synesthetic pathways nearby. It also made sense that there would be more noticeable activity in both the face recognitions and color areas of the brain because these areas are cross-activated. However, the amount of activity in my occipital system was unusual. Was the visual part of my brain just that more active in general? If so, was this due to biological reasons from an exposure in my life? Could it be due to genetics? Or, most likely, both? Was it neuroplasticity, the creation of new connections in my brain as a part of acquiring new skills? Maybe I had just developed the visual part of my brain from all the television and observation I did growing up, similar to how the areas of the brain involved in memory, language, and hand movements of accomplished lifetime musicians appear more developed than nonmusicians. Were there somehow just an increased number of connections in my occipital lobe?

David wanted to try and rig an impromptu system to measure the mirror-touch experience with the MEG, but because the discovery had just cropped up the day before, there had not been enough time to prepare. The best he could set up was a scenario in which he would scan my brain while I watched someone whip themselves with what looked like a fiber-optic horse tail. Needless to say, we scrapped the idea at its outset.

Without undergoing this final experiment, I returned to Iowa where I resumed my research under Peg's cheerful mentorship. Near the end of my fellowship I successfully put together a manuscript describing my study of the structural differences in parietal lobe development between males and females. At the time, there was evidence suggesting that men tended to score higher than women on visuospatial tasks such as mentally rotating a three-dimensional object. The lab previously found that these differences could be linked to measurable differences in the parietal lobe, the region of the brain responsible for visuospatial function: greater parietal lobe gray matter volume in women was disadvantageous on tasks of mental rotation while greater parietal lobe surface area in men was advantageous on the same tasks. Though I was curious about the results.

Even though a large portion of our brain's developmental changes—mostly the pruning of connections—occurred around the time of puberty, sex hormones did not seem to play a major role in contributing to differences in brain structure between males and females. Having observed mostly women with exceptional talent in visuospatial skills who were successful in engineering, biological sciences, and similar careers, I wondered whether sex differences in brain structure and function existed since childhood, possibly due to unchanging genetic causes, or if these differences tended to arise later in life, due more so to the influence of our physical and social environment. Examining brain MRIs of healthy people at all ages from childhood to adulthood, our results suggested that girls had greater parietal lobe surface area compared to boys but that around the time of adolescence, the difference flipped. Males then had greater parietal lobe surface area as they transitioned into adulthood. Taken together with existing evidence showing heritability of brain structure and the capacity to "train away" sex differences in visuospatial ability, our findings hinted toward a more complex and dynamic picture where one's changing environment and biology had an intertwining influence on the structure and function of our brains, and therefore, our realities.

With liberty in my schedule while conducting research, I simultaneously dug deeper into learning about leading groups. I got more involved with the American Medical Student Association (AMSA), a national organization of medical students. The efforts of the organization spread across my interests, from global health to health policy to gender and sexual diversity in medicine. Since my first year in medical school, I had steadily worked up the ranks in the organization, first locally, then regionally, and then nationally. I was delighted to be among others who were generally lighthearted and encouraged everyone to come as they are. I think it was AMSA that gave me my final boost through "gay puberty" to be more than just comfortable in my new skin, to feel whole with myself. By the time I was in Iowa, I had just been elected as the national vice president for the organization's membership, which allowed me to travel across the United States, attending conferences and meeting with other medical students. I felt immensely privileged to be able to soak in the personas, habits, strengths, and tenderness of so many others.

I did not want to prolong the already drawn-out traditional path of training to become a physician or clinician-scientist, but with all that I had been learning, I felt it was important to have an even stronger, broader base to continue honing, mastering, training, refining. I applied to the recently launched MD/MBA program at the University of Miami. The first two students from my original medical school class had just completed the MBA portion of their coursework. I saw the opportunity to become a part of the first full class of MD/MBA joint-degree students as serendipitous. I was accepted to the program and was thrilled, feeling like I was back to blazing a path wholly my own.

CHAPTER 4

Like Lazarus

On the morning of May 6, in the last weeks of the research fellowship, I woke up early to drop off my roommate at the airport. It was around 4 A.M. and dark out. The Iowa highways were typically deserted in twilight hours, which made it a rather straightforward drive. We spent most of the time laughing and lip-syncing to our favorite collection of pop music on the radio. When we pulled up to the airport, I wished her luck, then headed out. I turned the music back on and drove. I pulled up to the intersection of the airport's exit. That was the last clear memory I had that day.

I remember snippets of images. I remember driving down the highway. I remember setting the car on cruise control at seventy miles per hour, the speed limit. I remember being jarred with a loud thud under my feet. My headlights beamed at grass whipping underneath the car. I had drifted into the sunken median between the two highways. I reflexively turned the steering wheel right to get back on the road but not before my car came upon the culvert that highway patrolmen use to cross from one highway to the other. The incline acted as a ramp.

Nothing. I was once again in a place of nonexistence, just as I was during the surgery to remove my angiofibroma. I slowly realized that my mouth was filled with a gritty substance. I could not tell whether it was dirt, fragments of my teeth, or broken glass. My first coherent thought was automatic, mechanical. "What is my clinical status?" I could not see. I knew my eyes were open but blackness was all there was, not even contours of light or shadows. Just black. I performed a mental body scan by sensation. I felt a cacophonous array of pain everywhere but no sharp overwhelming pain in my head, my neck, my chest, or my abdomen. I could be hemorrhaging, but as far as I could tell, at least there were no obvious puncture wounds or large lacerations. I commanded my left arm to move, but it couldn't. Instead, I felt an excruciating amount of pain. My arm was either broken or dislocated. I did not know where I was or what happened. I had to cry out for help for someone to find me. I took in a burning, painful breath to yell, "Somebody please help! I've been in a car accident! Call 911!"

All that fell past my lips was a whisper.

I have no recollection of how much time passed while I mouthed for help. I could barely hear my words over the distorted garbling of colors from the radio, or from the pain, as if I were submerged in blackness mangled with distorted images from a broken television receiver.

"Are you okay?" came from the driver's side window. I replied, "I need a hospital . . . Also, turn . . . radio off."

My next memory was still in blackness, but this time I found that I could not open my eyes. It was probably for the better.

"Were you drinking?" came from my right.

"No."

"Were you using any drugs?"

"No."

I heard in the background, "Fortunately, he has a medical record with an emergency contact, but we've been calling and no answer."

I tried to interject on my behalf. "Page Nopoulos. Page Peg Nopoulos."

"Peg Nopoulos?" I heard typing from a keyboard nearby. "Psychiatry? Why would he want his psychiatrist?"

"She's not my psychiatrist. She's my research mentor . . . Please, page her."

From my left came another voice, "So you've dislocated your elbow here. I'm going to reduce it right now, so we're going to give you some meds for the pain."

I felt a cold rush come up my right arm through an IV, then I was gone once again.

I woke up slowly to the sound of a familiar tangerine "boop" sounding off out in a hallway. I was covered in several blankets. My chin rested on the hard plastic of a neck brace. To my right I felt the warm release of relief in recognizing a familiar face. Peg was seated in a wooden chair.

She told me about what happened. How there was someone driving about a half mile behind me that saw the lights of my car spinning over and over again. The police reported that, based on the site of the accident, after my car hit the culvert it propelled off of it like a ramp. My car was airborne for about one hundred feet and then rolled along the highway for about one hundred yards from where it hit the culvert. I must have fallen asleep. That was the only explanation that made sense. My car was totaled into a heap of broken glass and metal. Somehow the engineering of the car's frame absorbed the majority of the impact from the rolling.

While the car was lost along with my phone and everything else in the car, I only sustained contusions to my heart and the top portion of my lungs in addition to the dislocated left elbow and a fractured wrist bone on the left. The rest of my body was covered in thousands of lacerations from all the rolling. My body looked as if it had been placed inside a shoebox with broken glass, then shaken. All the blood vessels on the surface of my eyes ruptured, splattering around my cornea a solid cranberry red.

I remained in the surgical intensive care unit for a few days before I was transferred to the trauma ward. My father flew up from Miami to help care for me while Peg offered up her home for us. I was started on opiate medication for the pain, which left me feeling sedated and irritable with minimal control over much of anything going on in my mind.

I spent much of the next year trying to regain function of my left elbow. Because my arm had been in a sling, enough scar tissue formed to cause contractures. I could only move my elbow a few degrees up or down. In order to regain a full range of motion, I had to push against my flexed arm to tear the scar tissue little by little, day after day. Because I wanted to regain as much function back in my body as I could, I worked on rehabbing my arm everywhere I went. But I could not do it alone against a wall or lying down without the risk of dislocating the elbow once again. I recruited friends, even when traveling, to help me push through the pain, literally. The pain of tearing through innervated scar tissue every day for a year was agonizing but necessary. When I visited Lee, he was horrified as he watched my face contort with the pain radiating from my elbow through the rest of my body. I would be blinded by a mixture of red and black, purple at maximum pain thresholds. I instructed Lee to ignore me, to ignore my face or any sounds that I made and just push on. Watching Lee's calm face or the calm faces of my other friends helped minimize the pain. At least that way I was temporarily immersed in someone else's body rather than my own, which writhed in pain. Watching them, I happily floated somewhere between total embodiment and total disembodiment.

With additional occupational therapy in Miami, I regained most of the range of motion of my left arm. The orthopedist I saw in follow-up looked at the original X-ray before I came into the clinic room. It was an image of the tip of my humerus more than halfway past my radius and ulna toward my hand. He was incredulous that I was able to regain as much function of my arm as I had.

Eventually, my eyes cleared from cranberry red to pink to white. Scars formed all over my body. Most small. Some large. In particular, I had a new scar on the right side of my head that came off of my original surgical scar from the tumor. To me, the scar resembled part of a neuron. I embraced it and let it reinforce the original scar's message. It would serve to remind me of where I had been and where I was going throughout the rest of my medical training.

I was able to start business school as planned, finding immense delight in learning the skills and fundamental concepts behind why businesses, institutions, and organizational systems operate the way they do. I approached assigned business cases similarly to how I approached patients. Each case had its primary complaint and a history of present illnesses. There was a review of systems, a past medical history, a set of existing interventions, much like medications, and a contextual history such as family and social history. You performed your analysis like a physical exam and combined your findings with diagnostic data to formulate an assessment and plan. Thinking from the perspective of biological systems was exciting and different. I wanted to learn it all. Being that I learned best through practical application, I joined together with another medical student in the program, and we became the first medical students to ever compete in the university's annual business plan competition.

We proposed developing a mobile solution to fragmented patient medical records when people were still getting used to the idea of browsing the Internet on their smartphones. Competing against other more seasoned members of the business school, we were surprised and humbled that through some freakish luck we ended up winning the grand prize as well as the elevator pitch competition. Learning from all these new experiences and perspectives was the best form of rehabilitation for me after the devastating accident.

After business school classes finished, I was free to return as a full-time medical student to the wards. Because I had finished most of my required clerkships early on, I did not need to complete any more rotations after September. I was essentially free to do as I pleased until May graduation. I used the time to do some clerkships at other medical schools, called "away rotations," which are intended to give you practical ward experiences in areas that your medical school might not have available and to demonstrate your skills to residency programs you might want to apply to in the fall.

Applying was a prerequisite before you were invited to interview at the various programs that might be interested in you. Whichever residency programs you interviewed at, you could then rank them on your "match list." Each program, in turn, creates its own rank list. Both interviewees and residency programs submit their respective match lists to a national third-party program that then feeds the information into a computerized algorithm, which spits out your "matches" based on your mutual preferences. If you and the program rank each other as #1, you will likely be matched to that program. If the program ranks you low, but you rank them high, you could still be matched to their program if they had several available slots in their program. I was set on a neurology residency and wanted to have the liberty to rank whichever program I wanted highly with a strong chance of getting matched into it, so I participated in away rotations in Seattle and San Francisco. As a bonus, the rotations would double as a chance to explore the West Coast.

In Seattle, I spent two weeks on a consult-liaison psychiatry service where I had a consultation with a schizophrenic patient in the medical ICU. He had been admitted after attempting an overdose. Deaf, he communicated using sign language but had his hands restrained to the bed to keep him from harming himself, effectively leaving him unable to communicate. We made sure to provide a sign language interpreter and recommended adjustments to antipsychotic medications so he could have the restraints safely removed. I felt the movements of his body as he explained in sign language, "Do not restrain my wrists anymore, please." He pointed to the area where his wrists had begun to chafe. He also explained in the most feverish signing I have ever witnessed, "I was abducted by aliens in a UFO." He gyrated his hands in a halo, joining his index fingers and thumbs to show the UFO in flight. "I need to be able to communicate with them when they return, so do not restrain my wrists anymore." We were fortunately able to abide by his request, letting him know that we heard him, loud and clear.

My psychiatry rotation was followed by a month-long EEG and epilepsy rotation. The attendings on service thought that a month might be more than a medical student needed, but I disagreed. By the end of the rotation, I was able to gain an appreciation for the patterns of the squiggles moving beautifully across the screen. Their sharp, erratic movements had synesthetic sounds and were accompanied with reflected sensations of touch on my tongue and on my face. Like an EEG, the visual information imprinted itself on my own body map, lingering automatically in my brain. Groping at the squiggles with all my senses, I was able to accurately read and interpret simple EEGs on my own, a challenging feat even for most neurology residents because the majority of EEG experience occurs only if pursuing a dedicated epilepsy fellowship after residency. Having a basic level of EEG literacy helped me feel useful and closer to what it might be like as a neurologist.

I finished the rotation in late October with a strong sense of accomplishment. I wanted to go out and celebrate, to dive deep into my synesthetic associations, bathing and imbibing in music and lights. I arrived at a local bar far too early. Imagining that people would be out around 8 or 9 P.M., I walked onto an empty dance floor. I asked the bartender, and it turned out that most of the dancing actually happened after 11 P.M. To be exact, 11:30. I got a glass of water, found a couch nearby, and began to check emails when a small group of women at the edge of the bar asked me to come over and chat. They coaxed me over and introduced me to their friend. I had been lured into a trap. He was already rather drunk and insistent about buying me a steady flow of alcohol. I agreed to one drink and proceeded to nurse the sticky glass of rum and cola as if I were rationing the last glass of water in the Serengeti.

Being around even one intoxicated person, I begin to take on some of their disinhibited movements and characteristics through what I can only assume must be a form of top-down diminution of control. We chatted, though he was clearly becoming frustrated, resenting the fact that I was not guzzling alcohol to keep up with him.

The bartender was right. At 11:30, the dance floor filled with a throng of people. The music got louder. As I glided my eyes from person to person, experiencing the movements of dancing and new people, a set of hazel eyes caught me off guard, staring directly at me, unapologetically. I looked away, wondering if I had something on my face or my shirt. I looked back at this tall, handsome man. I walked toward him, slipping past elbows, shoulders, and high heels. I extended my hand, "Hi! My name is Joel. You're staring at me. What's your name?"

His baritone voice had the flavor and consistency of smooth, room-temperature Nutella. "I'm Jordan," he said. "Nice to meet you."

We hit it off that night. It seemed like we were both looking for the same kind of partner in a relationship. Up to that point, I had found myself almost exclusively attracted to people who were gorgeous blue 4s. Jordan was predominantly a reddish-orange 5 with some black 9, highlights of red 2 and, off in the background, a mix of turquoise 7s. He had only a hint of a few blue 4s. I figured that was probably good enough. Later in the evening, though, he became upset with a glitch on his phone, which he tried to fix by hurling it toward a wall. A committee of voices, composed of a sampling of all the rational cool-headed people I had met in my life, stepped forward and motioned to call a red flag. I promptly took the red flag, thanked the committee for their concern, and replied, "But his voice though . . . It's fine." There was that feeling of warmth and coolness that leaked from beneath his words as they slipped over me. The feeling overpowered all other rational thoughts. Barely noticeable, I swatted them aside. I placed the red flag in a mental bin and tucked it away, not knowing that one day I would run out of places to hide them all.

The following week I flew out to San Francisco to start my dedicated neurology subinternship where I filled the role on the wards just below a neurology resident. The patients we saw on the wards during that rotation were pathophysiologically diverse. Having spent most of my time on the stroke service back in

Miami, I felt like I was seeing the full breadth of neurology, from the nonphysiologic movements of functional neurologic disorders to the merciless jerking and spasticity of Creutzfeldt-Jakob disease. One of the patients I was assigned was a woman with transverse myelitis, inflammation of the spinal cord often due to an unidentifiable autoimmune or infectious process. She could not move or feel much from the waist down, but she was hopeful that she would continue to improve with treatments of intravenous immunoglobulin (IVIG), which helps block some of the body's autoimmune activity. She anticipated her upcoming discharge but was terrified of the idea that her insurance wouldn't cover her treatments. I spent several afternoons calling multiple insurance representatives until I was finally able to confirm approval for an appeal. She was so elated she burst into tears. On her last day on the ward, the senior resident supervising the team provided her a contact card in case she had any additional questions. She accepted the card and then looked my way and asked me for mine. She said she wanted me to continue to be her neurologist after she left the hospital. I was flattered that she thought I was anything more than a bottom-rung medical student. I thanked her for her kindness and told her she would have to be a little patient for that to happen—at least another four years until I would be done with neurology residency with fingers crossed that I might be accepted into a program of my choosing.

Another patient I cared for was described as confused by her husband before he anxiously brought her into the hospital for evaluation. She could converse with some fidelity of information but would quickly lose track of what she was saying. If someone spoke to her on the street for a few minutes, they might not think anything was amiss. Examining her, though, I felt the mirrored sensation coming from her—aimless, reflecting through me like an unexpected sensation of drifting. She appeared calm, but it was as though she were also lost in her surroundings, searching for where, exactly, to focus her attention. We ordered an MRI, which revealed that she had suffered several small embolic strokes throughout both of her cerebral hemispheres—thirty-two strokes, to be exact. We then ordered an echocardiogram, which demonstrated a clot on the heart valve closest to her brain, also known as a valvular vegetation. On the echocardiogram screen, the vegetation was visibly flopping up and down on the valve. With no sign of risk of bleeding on the previous MRI, we immediately started her on blood-thinning medication. An ultrasound study showed that about a dozen tiny clots would break off every minute and shoot up her neck vessels, putting her at risk for additional strokes. With no evidence of infection in her blood stream, the highest likelihood was a nonbacterial, or marantic, endocarditis. She had no abnormal clotting factors in her blood, so the next highest likelihood was that there was some kind of occult malignancy in her body that caused her blood to clot. We performed a PET scan of her body to see if there were any hyperactive tissue abnormalities that would take up more glucose than other parts, a suggestion of tumor growth. Most of the scan looked normal except for a little bit of increased uptake of the radiolabeled glucose in her chest. It was either right at the edge of her lung or tied to some active tissue in a cyst behind her breast. When she returned from the study, she seemed more preoccupied with the window of her room on the left. I wondered if she was captivated by something outside. I moved toward the window and she asked bemused, "Oh, oops. Where did you go?" We sent her for repeat imaging of her brain and discovered that she had suffered three new strokes, this time affecting the middle cerebral territory of her right hemisphere, which caused her to have a neglect syndrome.

Because we had to find an answer to the cause of her clotting, the following morning we arranged for her to have an ultrasound of the cyst behind her breast with the help of a pathologist to obtain a tissue sample. Ophthalmology came by to examine the back of her retina for evidence of any clots or abnormal vessels and also get a sense of whether she had lost her visual field on the left side in addition to the neglect. With the ophthalmologist, who turned out to be a former classmate of mine, we were able to confirm that our patient did not have a visual deficit or additional signs of clotting activity in her retina, which bode better for her recovery.

The following morning the woman was taken down for an ultrasound-guided fine-needle aspiration. Before she arrived back in her room, it was confirmed. She had an occult breast cancer hidden behind the cyst. This was the cause of her hypercoagulable state and therefore her marantic endocarditis. We made changes to her blood thinning medication and helped to get her breast cancer evaluated and start treatment. I was to leave San Francisco for Miami soon, so I stopped by her room before her discharge. She gave me her thanks for being there to help care for her and promised me that she would commemorate her momentous hospitalization by rechristening her husband's sailboat as *35 Strokes and I Don't Care.*

Outside of the hospital, in my final weeks of living in San Francisco, Jordan and I made plans for him to visit me on a few weekends. We grew closer. When I returned to Miami, I felt a longing to go back to Seattle and spend more time living in such a wonderful, progressive city while wondering if there was something between me and Jordan. To find out, and to further my career, I moved to Seattle, developing the concept and design on the mobile application I had proposed during business school while simultaneously figuring out what kind of life I wanted to live outside of work.

A few months later, as my match day approached, Jordan told

me he wanted to come with me to whichever city I ended up in. He and I both felt there was love between us. I was excited about the prospect but understood that this move would require me to come out to my family, my parents in particular.

I tried to script what I was going to say. I called friends and asked them how their experience had been with coming out. I researched online, reviewing videos and blogs and guides, trying to gain whatever additional tactic or strategy I could. I wanted to leave as little to chance as possible to be sure that my relationship with my parents would be reparable. I knew they would not be happy; that much was clear from derogatory comments they had made about gay people since I was a child.

I planned to have a friend from medical school wait in her car outside my parents' house. We were going to Key West afterward. She asked me what I thought would happen when I came out to them. I told her to have a box of tissues and a pair of dancing shoes on hand in the car, that I would use both, maybe one more than the other, but that was about all I could plan on. In the hour leading up to "the sit-down," my father was running late from an errand. I wanted to come out to both my mother and father at the same time. I wanted them to be there for each other. I asked if he could make it before I had to take off for my trip to Key West. He obliged and arrived.

My heart was beating through my chest. My palms were cold, clammy with sweat. I was quickly slipping away from my baseline regulation of sensory experiences. I sat my parents down on the couch across from me. My heart stopped beating.

"I have a boyfriend."

I felt the blood leave my mother's face. I felt my father's expression go blank and dissociate. He was no longer in the room. I started to lose my handle on my surroundings. I no longer felt my face. I was completely anesthetized, then light-headedness set in. As I felt myself slipping away, I let in a deep breath through my nostrils, hoping that the air would somehow buoy me for just a little bit longer. My mother went silent, trying to process what had just happened. I felt like I had betrayed them and, reflecting their emotions, that I had betrayed myself in the process. Why was I hurting them? Hurting myself? I was submerging deeper in their experience and losing my grip on my own. I had to hold firm like I do when I examine patients. I needed to consider my parents' feelings, but I also needed to ground myself. I concentrated on my toes, the feeling of my feet firmly planted on the ground. I curled my toes and felt them grasp at the fluffed cotton of my worn socks. I extended them gently, deliberately—flat, melding with the earth. I let in another deep breath, this time consciously allowing the air to reach the tips of my toes. The slow exhale helped me regain my physical and mental equilibrium, which helped keep me in my own experience while still engaging in theirs. I opened the floor to questions. My mother led the charge:

"What about Cristina?"

"I thought you liked girls, right?"

"What about children?"

"Don't you want to have children?"

"What do you mean you'll still be able to have children if you want to? How?!" She was becoming uneasy.

I considered the implications of my answer. "I'll tell you another day, but yes, it's possible."

I answered as many questions as I could, as many as I was willing. Once my mother felt satisfied, I realized that we had both been in tears the entire time. My father still had not returned to where he left, leaving his body hollow, occupying the blank space next to my mother. I told them we could talk later after I returned from the Keys.

Before leaving, I came out to my brother and sister. Neither flinched. They gave me a hug and told me that they loved me.

I used tissues and dancing shoes about evenly that weekend. I felt the guilt of causing distress to my parents but also had reassured myself enough up until that point knowing that I had to respect myself in the process. This was for them to figure out. I had to give them space and not resent them for anything that they would say or do as they attempted to lumber through unfamiliar territory.

A week later, my father drove me to the airport. "You know," he told me, "there is more to life than being happy." I could feel where he was coming from enough that, rather than being offended, I found I could simultaneously agree and disagree with each word of his statement. However, I chose to disagree. Maybe not a continuous existence of perpetual bliss and happiness; but what more was there to life than being *fulfilled*? I was grateful to have survived up until that point and was determined to continue living a fulfilling life.

For several months, my parents would continue to grieve the loss of their purely heterosexual son and the mirage of his future. I am not sure who my mother turned to during that time, but she sent me letters and emails asking me to consider conversion therapy or to consider my homosexuality a phase. She was torn. I chose not to take offense. It was almost always best to reply over the phone or over email, mainly so I would not lose myself in her experience and lose my own. That loss was too easy, especially with such an emotional charge. I needed to stay in my own skin, in my bones. I asserted to my parents that I was safe, that I was content. I would be there for them to help them through their process when they needed me, but that other than that, there would be no additional negotiations. This was my life.

To my surprise, when the week of the match arrived, my

parents agreed to have dinner with me and Jordan. To give them as much familiar ground as I could, I made reservations at an Argentine steakhouse nearby. The night could not have gone better. I was so thankful for Jordan's charm. He related with my father on sports, hunting, auto repair, and race cars, satisfying my father's insecurity about masculinity. At the same time, Jordan connected with my mother by telling stories about close ties to his family and Catholic roots. Jordan was atheist to be sure, but he had grown up Catholic, and that was probably most reassuring for my mother at the time.

The following day we gathered under a tent in the quad of the medical school. Name after name was called up to the stage. "Joel Salinas!" I stepped onto the stage. I fumbled with the envelope looking over the crowd, a sea of mixed emotions, blinking squirming faces, hair in eyes, tight neckties, and sliding shoulder straps. In front of such a large group, all the synesthetic sensations came together like television static. Noise. An occasional signal—movement or color or sound or feeling—might catch my attention from time to time, but mostly I was immersed in static. I loved being in front of a large group because, with enough static, I was simultaneously aware of myself and the crowd—a fragile resonance where, paradoxically, I felt most like myself. I took in a deep breath and pulled open the contents of the envelope. All I saw were colored letters scattered throughout the page. My adrenaline was running high enough that I had to take another breath to refocus and recognize words on the page. I announced into the microphone hoisted toward my face by the medical school's dean, a large flushed smile on my face, "Harvard Medical School; Massachusetts General Hospital for internal medicine internship, and the combined program of Massachusetts General Hospital and Brigham and Women's Hospital for neurology!" My first choice. Had I truly arrived onto that stage? Was I really saying those words? I paused for a second, wondering if the envelope was going to be snatched from my hands or if the words on the page would come into focus and reveal my real match. Neither occurred. I was grateful yet once again to the universe for affording me this moment. I savored it. I sat back down next to my parents. We kissed and hugged. I held Jordan's hand. We were Boston-bound.

ONCE I ARRIVED IN BOSTON, I WAS QUICKLY REMINDED THAT residency does not start at "residency." It starts with "internship," the first year of several years of training, where you are delivered unto the hospital ward with a freshly minted medical degree. If you are specializing in a surgical field, you are required to complete an internship in general surgery before moving to the anatomical destination of your choice. If you are specializing in a nonsurgical medical field, you complete an internship in general internal medicine. Sitting in an auditorium of more than sixty other medicine interns was electrifying. My knees quaked against the starting blocks, eager with anticipation to be on the ward and finally *officially* engage in the role known as "doctor." I wanted to learn as much internal medicine as I could so that when I transitioned into neurology the following year, I would feel competent in the parts of the body outside of the nervous system—the "supporting organs," from a neurologist's perspective. I fantasized being so assured with my knowledge of internal medicine by the end of my internship year that I would never have to consult an internal medicine specialist, at least not for the fundamentals.

Because I wasn't going to stick with internal medicine, I was cordoned off into a group known as "the prelims," a group of interns who were rarely brought into the fold or completely trusted in the innermost circle of other internists, especially during the first few weeks of internship. As such, prelims routinely drew the short straw, which was generally acceptable as long as the prelim's progression from rotation to rotation proceeded uninterruptedly. Rather than getting the anticipatory ward jitters over with on the first day, many prelims started with nonclinical electives. Electives were precious weeks with the potential to serve as a dire break from the wards, so you invariably wanted your electives to occur in the middle of winter when the nights were long and everyone's pleasant external demeanor had begun to rub raw.

My disappointment over not starting on a clinical rotation was short-lived, however. My first rotation was a self-design elective to learn about quality and safety in hospitals, and I made the most of the time. Secretly, I enjoyed it. The elective felt like a continuation of my business school education in understanding organizational structure. It afforded me a chance to learn the system at my new hospital; perhaps, I thought, it might even prove to be useful when I eventually did start my clinical rotations.

Because I had just arrived in Boston and was awaiting Jordan, I moved into a temporary sublet, a room in a small run-down apartment across the street from the hospital. Throughout the night, cascades of ambulance sirens passed outside my window, lighting my room and my mind with bright red before both settled back into the dusty grayish yellow from the nearby street lamp. A restaurant one floor below the unit closed down a few months before I moved in. Leftover food scraps made prime real estate for nesting vermin. Rodents often wandered their way into the apartment, a designated rat hostel. I was grateful for the short walk to work.

After two weeks of elective time, I began my first clinical rotation in a ward with private patient rooms. I was one of two interns with a senior resident tasked with supervising us both. The intern I was replacing was in his last week of his intern year. Handing over the responsibility of his patients to me, he seemed a little too deliriously giddy. He instructed me on what to do for each of the ten patients on the ward by reciting a rapid succession of bullet points and to-do lists, interspersed with glances at his watch and the occasional "Any questions?" Pale and light-headed, I kept my silence and politely nodded.

As soon as I signed the list of patients over to myself as the "responding clinician," a demon leapt into my pager. It shook violently with notification after notification. Possessed, the pager clung for dear life on my loose belt loop as it continued to go off. Something needed to be done for each patient, and I had yet to even meet them, let alone examine them.

Each day I found myself lagging, white-knuckling my way through the inefficiency of a cumbersome computer system, a cumbersome workflow, and most of all my own cumbersome inefficiencies, my own clumsy attempts at doctoring. I wanted to connect with my patients, to get to know them, to introduce myself as a physician they could rely on. But instead, I was firmly planted in a chair in front of a computer: entering orders; correcting previous orders; paging consultants; calling back consultants; taking orders from nurses; taking orders from multiple attendings, all of whom employed his or her own personal style. My patients were numbers and not the synesthetic numbers I had grown to associate with people; they were seven-digit medical record numbers, complete with their age, sex, room number, and a scribbled line or two about their disease, followed by column after column of checkboxes for the tasks that needed to be completed by the end of the day.

Just a few days in, I was already flailing in a thick hot broth of frustration. I questioned myself, "Doesn't anybody else see that these are people's lives? Why can't I be allowed the time to focus on one patient and on one issue, or am I only to juggle all of this just enough to make sure that no one goes into cardiac arrest?"

The patients felt that their medical team was not present except for their nurse and the attending, who only came by for a few minutes a day. I could not deny that I was not present. I was too busy being in several physical and mental places at once, which meant I was nowhere. I lagged far behind my cointern, who not only seemed to know exactly what to do in every situation but could easily finish every task ahead of time, well ahead of me. He also knew the jargon at the proficiency of a native speaker. He sounded confident, experienced. I unwittingly bastardized the medical vernacular and often blundered through clinical etiquette. Whenever I did catch up on my tasks, before I could congratulate myself, my senior or a nurse would approach me and attempt to kindly tell me how I had made an error and how my error, if not caught, would have killed someone.

I survived the first week, barely, and so did my patients, barely. The patients were never in any real danger given the multiple safeguards and checks in place to supervise fresh interns. But I still felt as if I could unintentionally hurt others. Entering the second week, I wondered whether I had made the right decision to go into medicine. I was unsure whether I had made anyone truly "better" by the time they left. When I finally began to pick up some of the efficiencies of the day, I took the opportunity to go and check on my patients for more than a few abbreviated minutes at the start of the morning. Ashley, a patient with cystic fibrosis in her twenties, had been in the hospital for several days, admitted yet again for yet another bout of multidrug-resistant pneumonia. Her sister was lying in the bed with her, flipping through a magazine. They appeared at home there, accustomed to the hospital. Ashley's infusion pump alarm had been ringing, and her nurse was busy with another patient. The sister grabbed my attention at the doorway, yelling, "Hey! You! Can you fix the IV?" I knew better than to touch the nurse's IV. As an intern, you are an easy target to be reprimanded passionately for existing—first by a nurse who's been traumatized by a succession of snippy faceless interns, then by everyone else in the hospital with an ID badge.

I tried to be honest. "Sorry, I'm unfamiliar with the IV, but I can get your nurse."

The sister tilted her head forward and raised her eyebrow, "Aren't you supposed to be the doctor? How can you not even know how to use an IV? You're *worthless*."

I walked out of the room, stung by her words in the full physical experience of their biting, objectifying tone. Her anger and resentment toward me mirrored in my skin, infiltrated muscle, touched bone. It only served to confirm what I had been thinking. If a patient tells me this, then it must be obvious how much of an imposter I truly am.

Across the hallway, I was entrusted with a patient, a longtime oncologist, who had been practicing for decades before I was born. A paraplegic, he was admitted to the service with recurrent kidney and bladder infections, despite having a suprapubic catheter burrowed through his lower abdomen and deposited in his bladder. He was retired, miserable, seething at all times. He spat vitriol at anyone who entered his room. On a few occasions, after watching me acquiescently nod at his diatribe about my assumed lack of intelligence and incompetence, he flung his toothbrush at me. Every morning, when I placed my stethoscope against his chest to listen to his heart, lungs, and abdomen, he pinched the front of my yellow oversized contact precaution gown, pulled me in closer, and imparted some words of wisdom. On the second morning of his admission, he asked me in cold whisper, "What are you doing this for anyway? Becoming a doctor? What a waste. I've been there. You just started. Get out while you still can."

I felt minimal movement from my midabdomen down, but the tension of his trembling shoulders carried twice its share of anger.

The feeling reflected in the lower half of my face gave away the secrets of the man's lax lower half. It echoed the dejection of an abhorred and abandoned shadow. I pressed my lips into my mouth and forced a half-smile, bringing the diaphragm of my stethoscope back onto his chest. "Okay, Dr. Harris, deep breaths in and out through your mouth." The instructions were more for me than him.

My attempts at trying to please everyone around me only succeeded in letting me achieve staggering breakthroughs in the field of utter failure. I could not make a single person happy without making others unhappy, or worse, lose trust in my ability. I would leave my apartment before sunrise and, after my shift, return to a hole with scurrying sounds echoing along the floorboards. Had I accomplished anything meaningful that day? My email inbox was peppered with LinkedIn notifications from former business school friends who had gone on into consulting jobs where they were taking on stimulating projects and making incomes 300 percent greater than mine. Uninspired, I was ready to take the good Dr. Harris's advice and quit. Why would anyone do this to themselves? I ruminated on the trope of a broken medical system, how I didn't want to be another piece of discarded tinder for the machine. All the value that I had hoped to bring was worthless.

I was ten days into being a doctor and drowning. I began to numb myself, letting all the synesthetic associations enter, come out of focus, blur into sensorial noise. I was swept by the undertow back into a numbness I knew well, far from waking sensate curiosity. I slowly sunk into a viscous haze. At least I was more efficient that way.

I started pacing the ward, checklist in hand, room by room, in submission, quietly counting down the hours before I could leave. I would dismantle myself into the cyborg needed to fill my assigned role, taking in the patient's emotion through my flesh and immediately dismissing it to move on to the next task. In an effort to eliminate my own pain and the pain of others, the simplest and most available solution was to muffle it by holding it under a steady stream of more—more patients, more checkboxes, more hours, more cordial, more perfect, more pain—to drown it out by drowning it. Under water, you can't fall. It feels secure, sheltered from all the hard and heavy edges of the external world. All your senses are both saturated and dulled. Living underwater lets you silence your eyes and slip away as the current gently pulls you under. To suffocate the senses, however, you must also suffocate that which sustains you, which keeps you alive, human. It is a false womb that comes at much too harsh a cost.

I sunk into a viscous haze. Four hours left. No. Three hours and fifty-nine minutes. My countdown was interrupted by one of the attendings. She came up and introduced herself over the counter in front of me, "Hi, I'm Susan Bennett. You're taking care of Mrs. Peterson, right?"

She must have seen my dim aimless gaze when I promptly replied, "Yes, what would you like me to order?"

She extended her hand in a greeting. She was the first attending to acknowledge me as enough a person to greet me with touch. I was somewhat suspicious, half expecting that she would pull her hand away in jest or fold four of her fingers to point at my face and reprimand me for having threatened her patient's well-being. But no, we shook hands.

"I'm Susan Bennett. It's nice to meet you."

"I'm Joel Salinas. It's nice to meet you, too . . . Is there anything in particular you would like me to order for your patient?"

She could sense that I was insisting on returning to the script. She looked over at my senior resident nearby. "Can you sign his pager over to you? I'm going to borrow him for a little bit."

I was taken aback but knew I could not refuse. After signing over my pager, I walked with Dr. Bennett to the hospital cafeteria where, after buying me a cup of coffee, she asked me how my week was going. I couldn't lie.

I told her how awful it had been, how I was having second thoughts about my career choice, how I felt as if I had been seduced into indentured servitude, bamboozled by an industry's branding strategy. She spoke calmly, a stark contrast to the constant feeling of crisis that I had begun to develop. She told me about her frustrations as an intern, the mistakes that she made, the regrets that she walked away with, how she continued on regardless, and how the first few months of our medical careers bear little to no resemblance to the rest of it. I found it hard to believe at the time, but I was so touched by her honesty. I felt purified by the reflected sensations of her serene demeanor. My shoulders lowered, my face lightened, and my muscles softened in what felt like the tender heat of an unexpected hot spring. There was gentle relief in a mysterious form of *surrender.* I had surrendered a flooded body and found myself in a new, more buoyant vessel. My breath returned, slow and secure.

The simple fact that a superior—someone who did not have to care, who could have simply given me additional orders and moved on with her day—not only took the time but made sure that I had protected time to reflect with her was more valuable than she probably knows. To this day, I still attribute that unexpected coffee break as one of the opportune moments that convinced me that it would be alright to remain in medicine, that most moments will not be perfect, and that finding the joy and gratitude in what I am doing will counterbalance all the other moments of crisis. Taking off the white coat and being more of a person and less of an institution was achievable. I just had to reprogram my brain's wiring to embody all of the physician reflexes and intuitions so that I could create the time I needed to speak with my patients, reflect, process, learn, enjoy, and be thankful. Real, physical, personal warmth is nothing like professional warmth. Personal warmth is kindled in the shared space between two people, in the willingness to be vulnerable *together,* to acknowledge each other, to sit in it exchanging breaths in the same air. Empathy is the hyperbaric oxygen needed to allow some of the most human wounds to heal.

Susan walked me back up to the ward. My senior signed my pager back over and, rather than feeling like a massive weight was thrust back on me, I took a deep breath. As I exhaled, I thought about how my rough start was not predictive of future failure. I was going to take issues one day at a time, and slowly I would make the strides I needed to finish strongly. By the time I finished my exhale, I felt the compressed onyx sheet in my chest begin to release and uncrumple, if only just a little.

Walking past Ashley's room I heard someone call, "Hey, you!" (my given name on the ward). Not being called by my actual name was just as well considering that the sound of "doctor" before my last name still sounded so foreign. I turned and saw Ashley sitting up on her bed beckoning me into her room. Earlier that morning, she had railed at me for asking her nurse to give her orange juice and a cafeteria frappe to bring up her low blood sugar instead of ordering a bolus of dextrose and several units of glucagon "like I always do!" Enraged, Ashley had yelled, "That's *malpractice*."

Now, Ashley sat cross-legged on her bed, wearing a white camisole and lavender pajama pants, which she had brought from home. She was alone. I stood at attention, unsure of what to do, what she wanted from me. Shadows from the afternoon light accentuated her gaunt, fragile frame. My cheeks and arms felt thin and bony. I was powerless, squirming in our shared frailty. I readied myself for whatever demeaning slurry of words she was preparing to hurl at me.

"I'm . . . I'm really sorry about what my sister said before."

"Oh, um, don't worry about it. The hospital's a pretty stressful place. We don't always say things that we mean."

"It was still pretty rude of us to treat you that way."

"Thanks. I appreciate your apology."

"Dr. Goleman came by. I told him what happened with my glucose. I tried to complain to him about you, but he told me that what you did was the right thing to do. I'm sorry I gave you such a hard time about it."

"No hard feelings." We shook hands. "I have to get back to work, but, if you need anything, just let me or your nurse know."

I walked out of the room feeling lighter yet slightly more planted on the rubbery hospital linoleum.

On one of the last days on the service, while auscultating the precordium of Dr. Harris, I realized that something seemed off. He was uncharacteristically docile. He had yet to pull me in for one of his dark pep talks. I glanced up at his face while listening to the late inspiratory crackles of his lungs, footprints of his chronic bedbound status. He was looking out the window. Sailboats quietly glided along the Charles River. He looked back at me as I removed the earbuds from my stethoscope.

"Have you read through all of my records yet?"

"Most of it, or at least whatever was relevant for your admission."

"So, you know how I ended up this way?"

"A gunshot wound."

"That's right. It went through my spinal cord. At the level of T9-T10. Did the notes say how it happened?"

I thought back to all the electronic notes I had poured through, most of which had been copied and pasted from the past medical history verbatim from one note to the next. "Actually, most of the notes just say 'paraplegic secondary to GSW,' but there's not much more detail than that."

"Do you want to know?"

"Only if you're comfortable with it."

"Of course, I'm comfortable with it. You gotta stop being so goddamn mousey and not be afraid to ask the patient the tough questions. That's where the answers are."

I nodded.

"I was shot in the parking lot of my clinic by a patient. He was pissed off at me for not being able to do more for his cancer. So, one night he waited for me to leave my clinic and shot me right in the back as I was getting in my car."

My heart sank, wondering if most patients hated their doctors.

"He was losing everything, so he decided to take everything away from me. He died a few weeks later. Meanwhile, my ass is covered in pressure ulcers and I'm pissing out a suprapubic tube. I'm still angry. Being here in the hospital, being an inpatient, it reminds me of where I was, what I lost, where I am now. Where I'm headed. Enjoy being a doctor while you can. It's actually not so bad. Most patients are just grateful that someone is thinking about them. Some patients can be real assholes, sure, but same goes for doctors and everyone else. So, is this your first year?"

"That's right. I'm an intern. This is my first rotation."

"Where'd you go to school?"

"Miami."

"Miami, huh? Boston's pretty cold. You sure you made the right choice?"

"Absolutely."

It was the first time I saw him smile. "I'll tell you what. Since this is your first rotation of your first year, I'll give you some advice from a seasoned expert."

He paused and pulled up the edge of his blanket and folded his hands together on his chest.

"You will never forget your first patients. Whether you want to or not, you will remember me. Remember that there is more behind a patient than just their medical record number. I didn't learn that lesson until it was too late."

He grinned. I felt his authenticity around my own eyes—the outer edges tightening gently, forehead remaining still. "You'll be just fine."

"It was a pleasure taking care of you, Dr. Harris."

He chortled in disbelief.

"Thank you for all your patience with me."

"You're welcome . . . Now get the hell out of my room."

At the end of my first ward service block, I felt that I could at least check off all of my to-do, "swing list" boxes by the end of the day without causing too much chaos. One step forward. I had an entirely new rotation the following day. I was coming onto Bigelow, a name used to refer to the hospital's inpatient general medicine wards. Each of the Bigelow services was labeled A through D. I was coming onto "Big C," not only the informal abbreviation for the service but also a rapper-like moniker that coincidentally alluded to its size. With more beds than Big A, Big C was the largest of all the general medicine wards. We worked in teams of four interns, supervised by two junior residents who were, in turn, supervised by two medicine attendings. Logistically, only two interns were present throughout the day, caring for somewhere between twenty-four and thirty-six patients at a time. Within ten seconds of walking onto the service, interns were sucked into a vortex of constant intensity. The list of patients consistently pushed the limits of what a single intern could possibly carry. The Bigelow pushed you to learn quickly and efficiently, although the start of each rotation was incredibly painful for everyone, especially for the nurses, patients, and the stooge-of-the-hour, the "responding clinician."

Meanwhile, the two junior residents were charged with being the team's taskmasters to ensure that the service ran smoothly. They were also meant to serve as examples for the interns, imparting pearls of clinical knowledge whenever possible. But not all junior residents were natural-born teachers, let alone leaders. Every junior resident got a turn leading the medical teams, regardless.

I felt pressured to perform. Work-hour regulations had just been enacted, and interns were not allowed to work more than sixteen hours straight, which was intended to help prevent the more typical twenty-four hours of shift-work plus an allowance of an additional four hours for administrative tasks to add up to twenty-eight, which often included a few more hours of wrap-up that brought the de facto total closer to thirty-six hours. Because the duty-hour regulations had been stressed so heavily, chief residents and graduate medical education leadership obsessed over making sure that everyone left on time, though this did not necessarily mean you were provided any more support to accomplish the strict objective. You might be offered assistance from a senior resident, for instance, as an implicit opportunity to demonstrate "weakness" by accepting said assistance or, the more common scenario, you just worked harder, faster, more perfectly—condensing thirty-six hours of work into sixteen hours of pure madness.

After a week of Big C, I had to ask my junior residents permission to leave the floor. Not because I was having any particular difficulty but because I had run out of clean clothes to wear. I had no access to a washer or a dryer, and the only laundromat in the area operated during regular business hours. Leaving home before 5:30 A.M., returning home after 9 P.M., you might convince yourself that you were living in a city of eternal darkness except for the few moments of sunlight you would take in as you power-walked past a hospital window to make it to the next patient. When it was your turn on the team to do a stretch of night shifts, you were dubbed the "nightfloat." You would care for the entire ward mostly by yourself, including any new admissions, with backup provided by a senior resident who was divided up among several interns in the hospital. You were encouraged to contact your senior. But there was an unspoken implication that contacting the senior meant you were "weak" and typically not be to be trusted or given more autonomy. We all wanted autonomy or at least to rise gradually in respect earned in order to make the cut above bottom-feeder.

My first night of nightfloat, I got caught up admitting five new patients, which at the time felt like an extreme number. I barely had time to check on the existing patients, let alone write notes and new orders and get pass-off information from a disgruntled emergency room resident or physician assistant or, god forbid, an emergency room attending ready to force-feed you whatever bits of information they knew about a patient in thirty seconds before turning their attention to another patient. One of the senior residents keeping tabs on my service saw the number of admissions I had and came over to check in on me. It was actually someone who had been a medical student with me during my first three years of medical school before I took time away for research. She was the senior supervising me and I was the intern. These gaps in hierarchy are well demarcated early in training. She was sympathetic and offered me some advice, "See one new patient, go to the next new one, and just keep moving. Notes can always be written later." Or, as the residency program director at the time would say, "Nobody ever died of a-note-emia."

Another senior resident, whom I had met outside of the hospital through mutual friends, courteously popped in to say hello.

While I sorted through a different patient's history to figure out how to prevent him from crashing, I noticed the senior resident pressing buttons and turning dials on one of the telemetry monitors that kept track of a patient's heart rhythms. Months later, he admitted that he was actually helping me out with a patient who, unbeknownst to me, had slipped into a sustained overly rapid heart rate. The telemetry alarm had been going off, and the senior resident did not want to rile me up any further. Remembering what was going through my mind that night, I thanked him profusely for making sure that the night did not become any more ludicrous than it already was.

On Big C, I was lucky to be graced with attendings who were able to balance patience and high expectations fairly. They were two strong women whose presence extended far beyond their physical height. They would pull each of the interns aside and force-feed their teachings to make sure that we were doing more than just mindlessly checking boxes. My first formal presentation of a patient on morning rounds was defiantly practical and abbreviated. "This is so and so, he is this many years old. He has Crohn's and he is here again with another flare. He is on bowel rest. The plan for him is bowel rest." In my sleep-deprived delirium, it seemed rational to leave 95 percent of the history as mystery. Both attendings took me aside at the end of rounds and gave delicate yet firm instructions that even if the night was impossible and even if there was a constant state of crisis, there are expectations to understand the bare minimum of a patient, their exam, and my own thoughtful assessment and plan. I had not met the bare minimum so much as I had flouted the bare minimum. I was ashamed and embarrassed. It was an unintentional disregard of *roundsmanship*—defiance through ignorant omission. I honestly assumed that I was providing the most succinct history possible and that my efficiency would be appreciated, if not celebrated.

I listened closely to the tag-teamed attending feedback. I took notes. I reviewed the notes at home. I added the feedback to my list of "areas of improvement," and I worked on each as diligently as I could. By the time I had to move on to Big A, on the last day of the Big C rotation, I was able to present a patient in the high flare and flourish of theatrical roundsmanship. Admittedly, not to full effect because I still needed notes in front of me for reference. But I lucked out by narrowly missing one of the new attendings on Big C who was notorious for snatching interns' patient notes out of their hands, insisting they recite as much of their patient's history from memory as they could. Instead, my merciful incumbent attendings were able to take me aside once again and congratulate my progress. One more step.

At times, the experience of being an intern felt so ludicrous that I was convinced that there had to be hidden cameras tucked away somewhere as part of a grand avant-garde social experiment. You could never openly mention any of these frustrations, of course. Complaining and whining had the potential to be considered a show of "weakness," proof that you were not to be trusted or rewarded. Despite the madness, it was undeniable that we were, in fact, learning. On one morning after finishing a grueling nightfloat shift, I made my way home but was so sleepy that I accidentally caught the wrong subway line. I went two stops in the opposite direction before realizing what had happened. I wandered through an unfamiliar station until I finally boarded the correct train. I stepped in and found a vacant pole to lean against, though the doors to the train were not closing. The passengers around were staring at me. Was it that I was wearing scrubs? Maybe I should have changed? Did I smell? I furtively sniffed the air twice for a rapid self-screen. The coarse red and orange smell of fresh vomit wafted into my nostrils. I turned around to my left and in one of the seats was a woman with disheveled hair, face-first into her purse, which she had vomited into. I crouched down and checked to make sure she had a pulse. She did. Check. I felt the expansion of her chest mirrored on my own in long, slow staccato. She was still breathing with an open airway. Check. I took out my smartphone's flashlight to see if her pupils were abnormal. They were small and constricted. Check. With those three data points, I was pretty sure she had likely overdosed on an opiate, possibly heroin. There were no pills or pill bottles nearby. Her left arm had what looked like a track mark. I sat there with my finger on her pulse and watched that her respiratory rate did not drop down dangerously until emergency medical services arrived. As they pulled her out of the train on a stretcher, I gave a brief story of my assessment to the paramedics. They gave her naloxone, an antidote for opiate overdose, and she was beginning to awaken. As serendipitous as it was to end up on that train when I did, I could not help but consider that perhaps I really was learning and making the transition to true physicianship. Perhaps I did have something to offer.

And yet, I was still learning the system and making medical etiquette errors such as presenting a *private* attending's patient to the *ward* attending. I concluded the presentation by putting the patient's acute illness in the context of his larger terminal illness. The patient's true attending had been avoiding the discussion about how the diagnosis was terminal for well over a year. Even though I committed such a blunder, I felt vindicated that the ward attending then took the time to break the terrible news to the patient. There was shock and there were tears, but at least we had helped to lay the groundwork for ongoing frank discussions about prognosis.

The weeks were passing quickly and I found myself steadily morphing into a full-fledged internist. I was efficient. My medical instincts grew sharp. I was better able to anticipate clinical decisions, wax poetic on common medical diagnoses and treatments, impress seniors, consultants, and attendings, and make the time to *sit and spend time* with patients and their families. I was even doing the unspeakable act of enjoying myself at work.

I made friends with some of the cointerns and, in particular, my coprelims. We cared for each other and looked out for each other. We shared in the war stories of being on the front lines, in the trenches. For instance, a reliable handful of patients on a general medicine ward were being treated for alcohol withdrawal. These were patients who suffered from multiple medical and psychiatric comorbidities, which were all exacerbated by their ongoing substance abuse typically involving alcohol. If they missed a drink or two, within about twenty-four to forty-eight hours, they would be at risk for delirium tremens, an acute state of mental confusion with tremulousness and increased probability of withdrawal seizures. These patients were often treated with an almost routine administration schedule of a medication to offset the withdrawal, known as benzodiazepines.

One night, I admitted a seasoned army veteran, Mr. Curtis, who turned out to be one of the most agitated cases of withdrawal I had ever treated. He required continually escalating doses of one of the strongest though short-acting benzodiazepines available, lorazepam. I felt the physical sensation of him spitting, swearing, and suddenly tugging against the leather wrist and ankle restraints tied to his bed—a violent, erratic, and jarring synesthetic chaos tugging at my body in the shadow of a dark hospital room. I was locked in phantom restraints. But I couldn't release him to free myself. He was at too high a risk of hurting himself and others. We had to reinstitute the use of phenobarbital, an older medication within a similar family to benzodiazepines. Mr. Curtis would be the first one on the general medicine services to undergo the "phenobarb protocol," which inspired anxiety in all the attendings, nurses, and pharmacists who associated phenobarbital with its other use, chemically induced coma. Mr. Curtis had been so agitated that at one point during the night he leapt out of his bed naked, cursing and throwing anything he could get his hands on. My cointern on overnight was in the call room on the ward where he was finishing up a patient note when he heard the commotion. He began to slowly open the door when a nurse screamed at him to "Shut the door!" just as Mr. Curtis rammed into it like a defensive lineman. Security was called. Tall, sturdy men in suits and earpieces rushed in to restrain him. Mr. Curtis was covered in antifungal powder and now so too were the black suits of the security guards with distinct powdered prints of hands, feet, and, in some places, traces of genitalia. We were skeptical of how the phenobarbital would work in such a challenging case, but we were awestruck the following morning. There was Mr. Curtis, sitting up in his bed in a neatly tied patient gown, hair parted to one side, listening to classical music and flipping through the morning paper. The concave chest and averting gazes I typically feel in the presence of guilt was completely absent. He had no recollection of what had occurred over the last forty-eight hours. From that day on, I swore by the power of phenobarbital for alcohol withdrawal.

As my intern year began to wind down, there was a secret thrill in realizing that as prelims we had racked up enough Bigelow time that many of us actually had grown to be more experienced on the general medicine wards than most of the junior residents supervising our teams. We certainly clocked more hours than the internal medicine interns who split their time between the ward and outpatient clinics. Even if it was virtually impossible for prelims to participate in "Reflection Rounds" to debrief on challenging moments of the intern year or attend most structured didactic sessions—because prelims were constantly at the beck and call of each ward—most if not all differences in education and ability that might have existed in the first few months of internship eventually faded with the practical experience of caring for patients.

While I was able to survive alongside my patients, there were two instances that shook me at my core. Margaret, an older woman with end-stage lung cancer, was admitted with pneumonia. Her code status was designated DNR/DNI (do not resuscitate/do not intubate). In the case of an emergency, her dying wish was not to have chest compressions or shocks administered to her heart nor to have a breathing tube slid down her throat and into her lungs. During a particularly frenetic day, a nurse rushed out of Margaret's room and yelled at one of my cointerns. Margaret was no longer breathing. In a panic, he rushed to Margaret's bedside then ran back out calling, "Code blue! Code blue!" We all ran into the room, the cluster of nurses and residents filing in as the intern started administering compressions. The code cart had just arrived when one of the nurses ran in telling everyone to stop. "She is DNR/DNI!" she yelled. "DNR/DNI! STOP!" We made Margaret comfortable and, shortly thereafter, her death was pronounced. I hoped that her death was painless, that she didn't feel herself suffocating. But I couldn't ignore the feeling of the compressions. Each one echoed in my own chest, each one a separate act of desperation against life's slow and sudden denouement. At that moment, the membrane between Margaret and everyone else in the room was torn open. The frenzy of compressions and shouting accentuated the ordinariness of Margaret's death. Her last moments, the final sensations she and I shared, were exactly what she had explicitly expressed she did not want. We all felt her death, everyone peering into the same hole Margaret left behind. Lingering in each of our eyes was the reflection of a dark abyss.

Through my own eyes, whenever a patient died, I felt as if I had died, too. The feeling never waned. In this regard, I have died many times. Watching patients pass away, I realized in my body the final moments before fading into death—the final seconds of futile panic or, more often, a silent act of grace. In these moments, I was always compelled to do something, anything, to try and prevent an undesired, accidentally traumatic and chaotic death. I wanted to honor a patient's final seconds of consciousness, to bear witness to the final seconds of his or her existence. Respecting a patient's dying wishes was worthy of ensuring zero flaws in the process because death does not arrive with the brio of lightning, a bright finale. It vanishes suddenly, like mist at dawn. It is all around us, dense, and fades just as we begin to look upon it with dignity, with reverence, leaving behind the distinct impression that we witnessed the final moments of one existence to remind us of our own. Living beyond death, we tighten our grip on it in one hand while drawing our other hand closer to hold a life we have yet to live. Experiencing another person's physical demise, I find myself in a sensorial moment of silence, a biological meditation before I once again inhabit my body, a cathedral without walls. Like Lazarus, I stand regularly at the threshold and behold in the distance an altar with enough space for a new sense of the divine and an even greater liberty to fill it once again, however I please.

CHAPTER 5

The Mind Has Mountains

EARLY IN MY THIRD YEAR OF MEDICAL SCHOOL WHILE ON clinical clerkships and before my year of research in Iowa, I was able to acknowledge the intense connections I was inadvertently forming with my patients. Immersing myself into a patient's experience was valuable. It was an opportunity to learn more about who they were as thinking, feeling, multidimensional people rather than the early academic reflex of imagining the patient as the subject of a scripted clinical vignette followed by multiple choice responses on a medical school exam.

With a few foundational clerkships under my belt, I was feeling a little bit more confident in my ability to interact with patients, gather basic information regarding their "history of present illness," and to begin translating it into medically relevant bullet points. I found myself spending a great deal of time on the history. It was almost impossible to resist the temptation to ask one more time, "And then what happened?" There was never enough time to sate my curiosity. I was fascinated with the story, and not so much my own interpretation of what the patient was describing as trying to understand what the patient was perceiving. Much like my need to grasp pathophysiologic concepts by building from the level of the atom all the way up to the social level, I needed to get a sense for how everything fit together, even as far back as their parents or grandparents, if necessary, to understand how such factors, whether emotional or genetic, helped shape the patient and brought him or her to me.

By the end of the internship, sorting through my synesthetic experiences in the hospital environment had become second nature. Though beyond the synesthesia, some of the admittedly distorted thoughts lingered. Whenever I caught myself immersing too far into the sensorial world of others, I reeled myself back into my own body. And, inversely, I came to see the process of evaluating another as delicate a process as surgery—done best with a steady, experienced hand. Developing the skill to objectively evaluate the self, to perform *self-surgery,* required even more deliberate practice. Taking the time to reflect on performance in a direct, authentic, and compassionate manner seemed to be the only way to learn every sinew, vessel, and nerve in my body. With deliberate practice, I focused on mastering the ability to traverse gracefully through my core—maximally effective, minimally invasive. It was hard to believe that I was able to move with that lightness, moving in and out of rooms on rounds while maintaining the flexibility that I did. That wasn't the case when I first started seeing patients as a medical student.

Before I started on the psychiatric ward, I had already begun to get a sense that specializing in the study of the brain was the right fit. To me, the brain was the motherboard of a person's reality. I found it inspiring that, though we might be able to transplant the heart, kidneys, and lungs, we might never transplant the brain. Even if we could one day, we most likely wouldn't elect to replace our brain with another's brain. Who would we be, with another person's brain in our head—a new person, a new entity entirely?

If I decided to pursue a brain-related medical specialty, my main options were either neurology or psychiatry, fields that were originally one until they explicitly split off into two after World War II. The divide had been gradual and began with artificial and staunchly held beliefs about the causes and mechanisms of disease when the tools to study their underlying pathophysiology were still developing. If there was identifiable damage to the brain, such as a tumor or a stroke, then the disease was more likely to land into the field of neurology. If not, psychiatry. Over the last half century, as the two fields continue to use similar diagnostic and treatment tools and operate under a similar theoretical framework, they have steadily approached something close to a reunion. Because both fields came from similar foundations, I was unsure about which was a better fit for me; they were equally enticing in their own idiosyncrasies. Psychiatry would be ideal because of its focus on understanding a patient's entire context, particularly social. But equally appealing was neurology's enthralling focus on neuroanatomical mechanisms. Its comprehensive methodical approach from the micro to the macro similarly suited me. I decided to use my clerkships in neurology and psychiatry as tests, starting with a psychiatry rotation, an ideal entryway into the brain and an opportunity to make a concrete decision. I was excited to begin a brain-related clerkship. I felt ready to go in and understand the patients' issues, to share in what they were understanding and feeling—in other words, to empathize with them and their conditions.

Yet whispers about how the inpatient psychiatric unit was even more mentally and emotionally intense than trauma surgery, coupled with horror stories of sick patients trapped and thrashing within the confines of their psychiatric illness, tempered my excitement. I was unsure how I would reflect disorganized and grossly psychotic patients. I considered the possibility of being tossed into perceptual chaos. I began my preparation for the rotation the weekend before, reading through clerkship guides to familiarize myself with the vocabulary of the disorganized mind. *Delusions of reference:* persistent crippling thoughts and convictions that others are talking about you, that the world is focused on you, and even the most common events have immense personal significance. *Thought broadcasting:* the delusion that others can hear or know your thoughts. *Thought insertion:* the delusion that others are putting thoughts into your mind. *Thought withdrawal:* the delusion that others are taking thoughts out of your mind. *Tangentiality:* where one continuously deviates from the main subject of his or her speaking. *Clanging:* the crippling association of word after word based on sound, without logical flow, impairing communication. *Word salad:* communication is unintelligible as seemingly random words spew in a mixed incoherent sequence.

Before I wrote off all my thoughts as ideas or delusions of reference, I was glad that one of our first lectures for the psychiatry clerkship was given by Dr. Steven Sevush, who had given the "split-brain" lecture that so captivated me during my second-year didactics. In this lecture, he introduced the theoretical *input-process-output* model for behavior. He stressed the importance of keeping a close eye on our own actions and inactions as clerks working with patients with psychiatric illnesses—just as much if not more so than their actions and inactions. He explained the intricacies of psychiatric disease in a language that made sense to me, understanding them foremost as disorders of the brain. In describing the neuroanatomy of psychosis, he used the example of *schizophrenia,* which is a term derived from two separate Greek roots: *skhizein* (to split) and *phrēn* (mind). These two terms were melded together through modern Latin to mean "split mind." The relationship does not imply the split of a person's personality so much as it refers to a person's mind splitting away from "reality." Once considered a single entity, schizophrenia is now thought to be one end of a spectrum. Schizophrenia spectrum disorders consist of observed symptoms of psychosis that typically emerge together. One component is disorganization—disorganized thinking, emotions, and behaviors that are inappropriate or unrelated to the situation at hand. The second component is altered perceptions that impair a person's grasp of reality, often referred to as psychotic symptoms, such as delusions and hallucinations. Disturbances in shifting attention lead to the inability to filter out information from within the brain or from the external world. Thoughts often coalesce as false narratives or delusions, which lead a person to believe that he or she is royalty or a messiah. A person can become paranoid that the world is about to succumb to a major cataclysm or that he or she is filled with parasites or is being persecuted by powerful unseen forces, just as he or she develops the obsessive conviction that a celebrity or public figure is in love with him or her.

Delusions are common in schizophrenia and can be composed of logical and illogical pieces of information sutured together that can be frightening, aggrandizing, or inconsequential. The capacity to combine unrelated thoughts and feelings into delusions in disease, when treated, is similar to the mechanism involved in generating creative thoughts that can lead to innovative contributions in a specific field. Sevush clarified such perceptual disturbances and hallucinations and other sensory experiences typically occur without any sensory input from the external world, which leads to a disturbed perception of what a patient is seeing or hearing. As a result, the patient is then unable to differentiate between what is occurring in their internal world and the external world, effectively losing his or her grasp of what is "real." The most common hallucinations schizophrenics experience are auditory, usually voices calling out their name or speaking with an abusive tone, co-opting the same neural machinery that most of us use for our own internal monologue. Disorganized and irrational behaviors and emotions tumble through the brain's circuitry, leading to laughing at a loved one's death or crying one moment, then suddenly becoming enraged before becoming cheerful, then terrified. Body movements, like thought and emotions, can be repetitive or absent for minutes to hours. Studies of the brain's chemical signals, cellular abnormalities, and shape of its structures all suggest that schizophrenia's symptoms tend to emerge from a disruption of how multiple neural networks interact and integrate, particularly the integration of multiple senses and sensitivity to stress. Neuroimaging studies specifically demonstrate that people with schizophrenia who are actively hallucinating show intense brain activity in the thalamus, the hub of the brain that relays and filters incoming sensory information. Paranoid symptoms have been linked to increased activity in the amygdala and other parts of the brain associated with fear. Studies of brain functioning throughout the frontal, temporal, limbic, and paralimbic regions also show that the disorganized symptoms of schizophrenia are tied to impairments and often to shrinkage over time in structures involved in behavioral, cognitive, and emotional regulation, including structures essential for filtering information of emotional importance, integration of senses, and judgment. Micro to macro, I felt this was a lens that could help give me a more even footing on the inpatient psychiatric unit.

On my first day I waited awkwardly in front of a heavy unmarked door with two other medical students. I bobbed quietly up and down, trying to balance myself on the outer edges of my worn dress shoes while thumbing through a series of notes fastened to my aluminum clipboard. We were in a building separate from the main hospital. Through the corridors video cameras nested silently in the corners, one conspicuously turned on the three of us. The sign above us read, in clean lettering, "Behavioral Unit."

We stood in front of the intercom and looked at each other to see who would ring the buzzer. I pressed and held the button for a second without a sound. The voice on the intercom crackled, "Hello?" I leaned in, unsure if we were expected or if the microphone was working, "Hi, um, we're the medical students? We're starting the clerkship today?" A loud buzzer—rough dark orange, mahogany, and black in my ears—sounded near the door. A metal lock clicked open. We walked into a brief vestibule. In front of us was another door dressed in warm wood with a thick glass porthole and steel reinforcements. A nurse at the station with thin frameless glasses sliding down the bottom of her nose waved her hand as she pressed another button. Another lock clicked open. We took a few steps into the ward, uncertain of what to expect. To our right was a nurse's station with reinforced glass windows all around. The entrance was set up like a waiting room with a television, table, chairs, puzzles, magazines. Deeper into the ward, a couch huddled with a few loveseats, and beyond them, chairs were arranged in a circle. Along the hallway to our left were doors to patient rooms, most of which were open and unattended with the exception of the first door directly across from the nurse's station. It was closed, its porthole covered.

We waited a few minutes until our senior residents arrived and walked us over to the physician workroom at the other end of the unit. This became a routine walk we made each morning—single-file, a half-stroll-half-scuttle past the patients. Taking that first walk, I recalled what I had taken from Sevush's lecture and attempted not to say or do anything that might accidentally give sensory input to a patient's disordered mental process. I looked straight ahead at the workroom door, avoiding eye contact, limiting my facial expressions, offering nothing outwardly to grab a patient's attention. Out of the corner of my eye, I noticed the patients in the back of the room pacing in circles near the chairs. Some were wringing their hands as if anxiously waiting for terrible news; some were staring at the ground as if in penitence. A few patients sat on the couch. One stared off into the distance with a yellow legal-rule paper in her lap, a squat pen resting in hand. The page was blank. Another sat in front of a puzzle on the coffee table digging patiently through a box filled with pieces.

We gathered in a cramped workroom without a porthole on the door. A long table was crammed in the middle of the room leaving just enough space for a few chairs around and a chalkboard. After about thirty minutes of labored introductions, the attending psychiatrist walked in. He introduced himself then proceeded with a chalk talk. He mentioned how some of the patients we may have already seen on the ward have predominantly "negative symptoms" of schizophrenia. "Most of them are in their rooms or out but don't do much," he assured us before telling us about the patients with "positive symptoms," the ones with psychosis and hallucinations. Later that afternoon one of the residents was going to present a patient admitted just yesterday. "This," the attending promised, "will be a good example of positive symptoms."

The resident went through the patient's story, though he spoke so quickly that I had a hard time following all the details. He rattled off a list of positive or negative symptoms before offering a frank description of the patient's behavior that felt to me as if he had gone off script. When he finished, a nurse brought the patient into the room and sat her down in the empty chair at the other end of the table. The resident had described a middle-aged woman with type I bipolar disorder, meaning—by definition—that she had at least one manic episode in her life, regardless of whether she also had a history of depression. In her case, she had experienced many manic episodes, often with psychotic symptoms. As soon as she recognized that her usual manic symptoms had begun, she brought herself to the psychiatric emergency room.

"Lord Jesus brought me in a time machine to stop you all from setting off the bomb!" she said after a few minutes.

Leaning back in his chair with his arms crossed, the attending asked, "What's your name?"

"My name is Naomi Cleopatra the Fifth, Goddess of the Nile, and all of Egypt and America!"

I was taken back by the explosive nature of her replies. I tried to understand where the fragments of speech were coming from. Was she trying to communicate that she was lucky that she was able to come to the hospital yesterday before her symptoms got worse? Or, were these remnants of memories caught up in the storm of her thought processes? I felt the sensation of her muscles tensing, her shoulders coming forward, her hands gripping the table before she spoke. I felt as if I weren't careful, I might accidentally jump up and exclaim an answer. My mirrored sensations were erratic and punctuated. The reflected actions of her body and thoughts seemed to curl and burrow through the air in different directions as if they were triggered mousetraps, springing-out-of-nowhere emotional reactions with yelling followed by a sudden drop into darkness.

"Do you have any children?"

"I have two-two-two hundred thousand, five hundred and thirty-nine children, all comin' out of me! All born in the year two hundred!"

The interview continued like this until the nurse eventually escorted the patient back out of the room. She would have her mood-stabilizing medication increased with an additional antipsychotic medication while she was an inpatient. I was like a spring under pressure. My thoughts continued to assemble the fragments she had laid on the table like a mosaic, trying to understand if there was meaning in her words, if there was an anagrammed narrative hidden among the discarded pieces. I needed to come back to a solid baseline. I stared at the doorknob, focused intensely on its curvature, how the light reflected off its metal surface. I felt as though my face had taken on the shape and texture of it, which somehow brought me to a calmer plane. My body began to relax; my thoughts steadied.

Following the patient's departure, the attending used the opportunity to give a follow-up chalk talk about psychotropics, medications used in psychiatry that have an effect on the brain. He sketched a picture of the brain on the chalkboard while describing how post-mortem studies have shown that patients with schizophrenia tend to have extra receptors for dopamine, one of the chemicals released in synapses that allows neurons to signal each other and is especially involved in the brain's movement, learning, attention, emotion, and the reward-reinforcement systems. The increase in receptors for dopamine possibly makes the dopamine systems more and more active, which results in an increased risk of hallucinations and other positive symptoms until a patient loses all capacity to sort internal and external stimuli. Treatment of psychotic symptoms involves a class of medications known as antipsychotics, which come in the form of "typical" and "atypical," though the mechanism is generally the same: decrease the activity of dopamine. These medications block the receptors that normally receive dopamine molecules in neural networks that control the initiation of motor and nonmotor actions, including thoughts. Though antipsychotics are typically the most potent, antianxiety medications, like benzodiazepines, sometimes help manage agitated thoughts and symptoms of agitation while other medications can stabilize the fluctuation between different mood states. When the patients are admitted to a psychiatric ward, the team works on getting to know them and finding the right medication or the right combination of medications.

The attending also stressed that one of the issues with the dopamine-blocking medications is that while they decrease psychotic thoughts and impulses, they also have side effects similar to what is seen in movement disorders like Parkinson's disease because they interfere with some of the brain's automatic mechanisms of movement. I experienced these side effects firsthand later when I interviewed one patient before admitting him to the ward. He was pacing as if he were on an invisible race track going in laps over and over again, hurrying to cross the finish line. He sat down next to me, and I immediately felt the sensation of nearly continuous movement around my nose and lips, sparked by the patient's form of fidgeting often known as *akathisia,* increased motor movements of his face and mouth, like a rabbit twitching its nose. Interviewing him, I had a hard time keeping focused on the conversation as I struggled to stop my face from mirroring the man's twitching, tweaking, pouting, crinkling movements. This was one of the instances when I realized the extent of my permeability, how quickly I can take on more erratic movements. The more spontaneous they are, the more likely I am to mirror them. The sensation on my face made me uncomfortable. I felt malleable, like wet clay being passed along a crowd of hands. Despite this feeling, I wanted to learn. I wanted to be helpful. I needed to get over my discomfort.

We had a Miami thunderstorm one morning. The attending psychiatrist walked in later than usual onto the ward, his Italian shoes squeaking, umbrella in hand. To make up for lost time, we immediately began rounding. He marched ahead of the group from room to room with a large, wrapped, golf umbrella in hand, asking the patients pointed questions. I wasn't sure whether his style was overly callous or whether his Spartan back-and-forth with patients was a symptom of someone who had spent so much time working with people with severe psychiatric illnesses, a form of coping on his part to shut out the patients. We barged into a dark room. He turned the light on. We were rounding on a patient, a woman in her midsixties, who was admitted with refractory depression and active suicidal ideation. She had just started electroconvulsive therapy (ECT), a treatment with an unfortunate name and even more unfortunate depiction in movies. ECT passes an electrical current through the brain while the patient is completely anesthetized with the guidance of an anesthesiologist. ECT is also one of the few treatments that has been shown to be effective in patients with severe depression who have run out of options and are at a high risk of committing suicide. The attending psychiatrist tapped on the side of the woman's bed with his umbrella as she stirred awake, pulling the blankets down from over her head slowly. He asked her how she felt. She paused, combed back a few strands of white hair, and replied with some excitement that she had just received her first treatment the day before and she felt that maybe she was already feeling less depressed. He asked if she was still having thoughts of killing herself. She paused again and said no, she didn't think so. He let her know that he was glad and he walked out of the room. Once we all emptied out of the room behind him, he closed the door. Looking away, addressing no one in particular, he mentioned slightly under his breath at a volume we could barely hear, "She's lying." He sighed and mentioned that it was too soon after ECT for her to show any improvement, that she might just be saying what she thought we wanted to hear so she could be discharged. "That's when patients like her actually do it. Kill themselves." He kept marching forward. I had felt the simultaneous physical sensation of carrying my center of gravity in my chest while also reflecting barely any change in emotional expression—a combination of detachment tinged with an unavoidable pain of grief and frustration and helplessness. When did that shift happen in his career? What brought him into psychiatry? How many patients had he had who committed suicide after discharge? Had he known someone personally who killed themselves? Had he struggled with depression? It wasn't my place to ask him any of these questions. But if he had walked out of some wreckage as detached as he was, then perhaps, as a fresh clerk, I could contribute to the care of these patients by deliberately showing more empathy here. I still thought of myself as an incapable medical student, mostly a nuisance to the scarred veterans of war. But I considered that this could be a chance to learn more about understanding where people were coming from, all of them, or at least begin to dispense with some of this hope that seemed to mark me as an obvious novice.

Back in the workroom, just as we began our usual review of who was planned for discharge, we heard a sudden muffled crash like wood slamming against cement. There was movement against the door, followed by an overhead call for a "Code blue, behavioral unit, code blue, behavioral unit." A security guard opened the workroom door and asked the attending to step out. A patient, standing in the middle of the encircled chairs, wielded a chair in his left hand, grabbing it by the leg as if it were a valise. He was enraged, aggravated. "No! I don't want no olanzapine," he kept yelling. "I hate that med!" He was over six feet tall. Gray sweats hung on his muscular build. I felt the heaving of his chest mirrored on my own, his furrowed brows, his knees bent, one foot slightly in front of the other, right hand in a fist and cocked back slightly, ready to strike. The security guards approached carefully. A small group of people dressed in civilian clothes stormed in walking hurriedly. Code blue in the psychiatric hospital was different compared to the main hospital. Code blue was a signifier of a behavioral event, usually someone becoming agitated or combative. The code team here was a group of people designated to come to the location where it was occurring and intervene simply by their presence. To give a sense that there was a larger presence of people could help de-escalate someone from violent behavior without using force and, with enough redirection, the patient could have a medication administered as needed to help him or her regain control. This patient was due for his antipsychotic and had become angered when he learned of the medication. He knew the medication had weight gain as a side effect, and he was fastidious about his weight. He quickly lost control of his anger, demanding, "I want the one that won't make me fat! I want the 'ziprasidon' one!" The security guard and attending let him know that they wouldn't have a problem changing his medication, but he had to calm down first and take the medication. I felt his breathing suddenly change with a deep inhale. His brow remained furrowed, but the fist in his right hand began to soften. He lowered the chair, and then sat down in it, his legs spread open. He leaned back, waiting. Because of the agitation, after receiving a dose of the ziprasidone, he was moved to the isolation room, the room across from the nurse's station. It was a simple room with nothing that could be picked up or thrown. He was confined in that room the remainder of the time I was on the clerkship. He was at risk of harming himself but, even having witnessed the incident, by then I had become much less frightened of patients with psychotic illness. Sure, in the case of an episode of agitation, backup was critical. But more often than not, the violence was not directed at you. More often, they were more terrified of you, terrified of what was clanking and clanging in their mind, terrified of themselves, the world, the uncertainty of it, pain. This wasn't just my isolated experience in the clerkship. The depiction of people that have a mental illness as violent sadistic people could not be further from the truth. The evidence from research firmly indicates that having mental illness is not a predictor of violence toward others and that, in fact, *they* are more likely to be victims of violence than victimizing another. The incident with the man encouraged me to challenge whatever prejudices I might have unconsciously cultivated. Maybe if I drew myself even closer, I remember thinking, I could discover more commonalities than I might expect.

A few days later I sat down with a patient admitted for a manic episode. Of the patients who were admitted with psychosis, the larger proportion were patients whose psychosis was sparked by the manic phase of their bipolar disorder. Believe it or not, bipolar disorder is quite common (affecting almost six million adult Americans) and can be found in successful professionals who, when they become severely manic, will either admit themselves or a family member will admit them to a psychiatric hospital. After a few days of having antipsychotics adjusted to help them through the most challenging part of the mania, they may be able to return to work. It's not all that different from a person who needs to be admitted for an occasional flare-up of an inflammatory disease. The patient I sat down with was a chef. In his mania he would write, a symptom known as hypergraphia. The majority of his writing was poetry. He stayed up the night before writing and came up to the nurse's station declaring that he had written a masterpiece. He wanted someone to read it. I stepped out of the nurse's station and asked if he would show it to me. Standing next to him I had the physical sensation, as if I had just drunk several shots of espresso. The feelings of mirrored movements were brief, sharp, frenetic, full of energy, bursting forth. I tried to think of a previous time I might have felt like that. On the first day of medical school, I ordered a cup of coffee at a Cuban coffee shop. By the late morning I was trying to claw my way out of my own skin, which was when I found out what I thought I had ordered, a *cortadito,* or a shot of Cuban espresso, was actually a *colada,* a cup of coffee so strong that it's meant to be split among ten people in thimble-sized cups.

The chef's poetry was effusive with supercilious, polysyllabic Latinate words heavily focused on religious themes and demons. The words circled around the subject of superiority, of knowing a truth that others, not even the world nor God knew or understood. Grandiosity shot up in mania. Though at the same time I wondered about how much art, how much music, comes from this place. I started to think about how often *I* might have had thoughts or feelings like that. To squeeze in one or two more hours of studying, I often skipped some sleep, sometimes for nights in a row. Perhaps I could be just as manic as this man. Perhaps we could all have some of these features that, if we looked within, we could easily discover in ourselves. I wondered how different his depressive phase must look. The man was not much interested to hear my opinion on his poetry. He collected it with other scraps of paper that were crumpled together and scribbled all over in blue and black pen ink and pencil graphite. He hurried back into his room to work on his masterpiece.

I wanted to continue diving down further into my commonality with the patients, to see if deep down alongside their psychosis I could find an insight or catch a minor glimpse of what was going on internally and what I could say or do to help provide some relief. A college student in her early twenties was admitted for psychotic symptoms. She was at the county fair with friends. They thought she was in an exceptionally good mood that day, but this changed when she started going on about the people in the crowd. They were talking about her. She could overhear them saying how funny she was, that most of them were in on her jokes. She kept on laughing. Her jokes became a little more inappropriate. She was beginning to embarrass her group of friends.

When they dropped her off at home, her behavior continued. Her mother became angry, telling her to stop fooling around while uncomfortably surprised that her daughter, a well-behaved high school honor student, could act out this way. She didn't sleep. Her mother's anger quickly turned to fear. She brought her to the emergency room where she was admitted for a first presentation of psychosis. The question had been raised whether her use of a steroid inhaler, which she had just started using for her asthma, could have led to a steroid-induced psychosis because this class of medications can, in rare circumstances, lead to symptoms of psychosis. The psychiatrist cut off the medical student presenting her case. "That's bullshit. They always blame it on steroid use, but this is probably a first presentation of schizophrenia."

Later that day I sat down with the girl to get a sense of what had been going on in the weeks leading up to her presentation to the hospital. To understand her behavior, I made a deliberate attempt to dive in. I asked her for her name. Giggling, she rattled off the names of cartoon characters.

"Where do you live?"

She started, "I live at one-sixty-five . . ." then paused, derailed by a thought. She contorted her face. I could see her struggle to stay on the tracks. "I live under the sea. You wanna come? That's why you're here right? Wait. I'm sorry. What did you ask?" She yawned. "I just wanna go home . . . But the Ninja Turtles won't let me."

Negative symptoms are absence of movement or expression, neglect of emotion, or withdrawal. Hallucinations and delusions are considered positive symptoms. Jumbles of thought and speech as she exhibited are the third category of psychotic symptoms. It was heartbreaking to watch, to feel, the young woman come up for air briefly before going back down. I could see her connect a few dots, then immediately, as a thought zoomed by, her attention would latch on to that passing thought and bring her as far as it could before she could break free and bring herself back. I thought back to my most vivid sensory experiences that once distracted me and drew me further inward. The young woman's experience was far worse. She could not keep a handle on anything. Her filter to separate her thoughts had been broken.

When I was a child I developed a fever and experienced what I could only describe as hallucinations. I saw the cartoon characters from *A Pup Named Scooby-Doo.* They were in my head, but I was so far in my head that they might as well have been in front of me. I watched them talk to me, in stuttering starts, freezing midway through sentences as they glided down a slanted hill. Interspersed with these characters was a giant, blonde, plastic Polly Pocket doll, which would zoom in, then zoom out to the size of an ant, back and forth. I could not make sense of these visions. I could not control them. Once my fever broke, these strange sounds ceased; the visions disappeared. The young woman was going through something similar. As I tried to follow along, I felt that I was sinking deeper into a depth of loose associations when the senior resident put his hand on my shoulder, startling me back into everyone else's reality. That I had so easily lost track of my surroundings was frightening.

After about two days of antipsychotics, the young woman seemed to be able to float toward the surface a little more easily, but she was still struggling. Three days after her admission, I overheard a nurse telling the senior resident that the young woman had "made a mess last night." According to the nurse, the young woman had wrapped towels around her head then stuffed the toilet, sink, and floor drains with paper towels. She turned on the faucet and let the water run over, flooding the bathroom. When the overnight nurse tried to put her back to bed, the young woman became agitated and tried to claw at the nurse. The young woman received a few additional doses of antipsychotics. By the time we were rounding on her, she was calmly sitting up in her bed. She had just received an increased dose of antipsychotics. She appeared different. To my relief, my reflected physical sensations were easier to predict. I could follow the logic of her words and her emotions, though her facial expressions, while still animated, were blunted on my face. I recognized in them fear, which presented itself as a dark blueish-purplish cowering heap in me. We asked if she recalled what had happened the night before. She said that all she could remember was that she had been invited to a house party with SpongeBob SquarePants. She raised her eyebrows as a form of admission that this sounded strange. She remembered trying to get the party going by putting the house underwater. Her instinct in the depth of her psychosis, as I recognized all too clearly, was to go deeper.

She and I were only a few years apart in age, the right range for new-onset schizophrenia. I worried that I could become like her. What if my synesthesia was not synesthesia at all? What if it was all the beginning of schizophrenia? These same thoughts surfaced again during my psychiatry consultation service in Seattle where I met a male patient similar to the young woman. He was in his late twenties. He had attended a pirate-themed fair. He and his fiancée went in full costume and were having a blast when, over the course of the day, he became more disorganized in his thinking. His fiancée was having a harder time following the flow of their conversation. They went home early. Later that night his fiancée woke up to an empty bed. After searching the house, she found him in the backyard, just standing there. She called out to him. He didn't respond. She called out again, and he began to take a staggered step forward as he raised his hands, which the finacée realized were covered in blood. A mess of lacerations ran along the length of his forearms. She frantically called an ambulance, and it was only later discovered that he had entered the woods behind their house and attempted to break the window of an old rundown pickup truck. Before this episode he hadn't exhibited any symptoms. He and the girl on the psychiatric ward in Miami both experienced triggering events for what was likely a preexisting substrate for schizophrenia, part of what is considered consistent with the "diathesis-stress model," which proposes an underlying *diathesis*—predisposition or vulnerabilities for schizophrenia, including genetic and other predominantly neurodevelopmental biological factors—and stressors such as adverse childhood events and extreme moments of crisis. Some people may possess within them a biological or genetic vulnerability to schizophrenia but, without sufficient stressors, never develop schizophrenia. This may also explain why schizophrenia tends to be higher in people who experience extreme or chronic poverty and socioeconomic stress—a scenario that is far more likely than the presumption that mental illness leads to the poverty and socioeconomic stress.

I realized once again how easily anyone can slip into schizophrenia. How easily this could be me. Though the odds of developing schizophrenia are about one in a hundred, these odds are four-fold greater if you have a second-degree member of your family with the disorder. The odds jump to one in ten if an immediate family member is affected with schizophrenia, one in two if an identical twin develops schizophrenia, even if the twins are raised apart. Because my age automatically made me susceptible to new-onset schizophrenia, I worried that continued prolonged and unadulterated exposure to triggering experiences put me at greater risk of developing schizophrenia. I was in the right age range, and I imagined that with enough triggers I, too, could be mentally derailed, leaving the wake of stigma and upheaval in my family. Most chronic illnesses are similar in their personal significance. But in chronic illnesses that don't affect your ability to think and feel and act, you have the wherewithal to reflect on your illness. In these cases, the patients had neither the insight nor the attention to follow through on what was occurring to them. I wasn't sure whether this scenario was better or worse.

I recall monitoring another patient on the consult service in Seattle. He had been locked in a room for weeks. I stood in his room with the psychiatry consult attending. The man appeared calm, quietly working on a crossword puzzle in his bed. While the attending spoke, I dove into the man's experience, drawing as much attention as I could to try to understand him, to empathize with him. Underneath a general sense of calmness, I felt around my eyeballs an acute uneasiness. The man's eyes darted from person to person, then to the table, back to me and the attending, then back to the table. I looked closely at the table next to his bed. A stack of books and that morning's newspaper laid on top. I noted the spines of the books: *Introduction to Electrical Work, The Study of Radiofrequencies, Circuits, Espionage, Wiretapping.*

"How are you doing today?" the attending asked.

"Oh, much better. Definitely. Can I go home soon?"

"We need to observe you for at least a few more days to see how stable you'll be on the current dose."

"Okay! Thank you!" and he went back into the crossword puzzle.

When we stepped out, I wasn't sure what to believe. I thought about mentioning what I saw on the table to the attending, the concern that he actually wasn't improving. I could hear and feel in my throat the angst in the man's voice. The desire to be somewhere else, to not be sick. Of course, I could not speak for him. But trying to understand and feel what he was going through was my compass. I mentioned my reservations to the attending. She stopped writing, put her pen down, and we went back in the room. She asked again about whether he was having trouble controlling any thoughts. He hesitated, then said that he wasn't losing control of his thoughts, but he was convinced his room was probably wiretapped, and he desperately needed to figure out who was watching him. To break out of his delusion, he was reading up on espionage and electronic eavesdropping. The attending increased the man's medication and kept him for further observation.

Observing the young woman in Miami, I felt this same pang of confusion. Though I believed I had started to understand her and what she was going through, I wondered if I even had the right, let alone the capacity, to speak and think as if I understood her. I was just a medical student, a sensorily overwhelmed one at that. I began to wonder if I had a personality disorder myself, wondering each day if tomorrow would be the day that I would be admitted against my wishes for a psychiatric decompensation. I wanted to be able to empathize deeply, to submerge in the mirror-touch sensations to join alongside them. But I began to fear whether I had the ability to return.

That afternoon I joined my senior resident in a court hearing to determine whether a patient had the capacity to make decisions on his own behalf and provide a court-appointed guardian. The courtroom was a small, crowded room within the hospital building. The judge sat at the head of a long table. The patient, who remained quiet, sat next to other county officials while the attending sat next to the judge. The attorneys reviewed the case with additional input from the attending who excerpted descriptions of the patient's psychotic and paranoid behavior—how there had been adjustments to the medication, how he had been a difficult case, and how there had been limited but steady improvement. After about twenty minutes, the judge paused. He turned to the patient and asked a set of questions that sounded as though he had asked the same questions a thousand times before. The judge asked the patient if he would be able to find food and shelter if he left the hospital. The patient began to describe what he might do. He suggested he could go to the local shelter. But then he went into his case, point-by-point, how he was in the psychiatric unit because there was a conspiracy against him by the government, how he had been set up by his ex-wife, how he was the victim. The court ruled that he needed a guardian, and he could continue treatment as an inpatient. The room felt cold, absent of the warmth of compassion.

The psychiatric emergency room service, also known in some hospitals as the acute psychiatric service, was an even more intimidating place. Here I helped evaluate patients who were often unmedicated, at the peak of their psychosis. I met with my assigned patients—as long as they were not too agitated—in small rooms, one at a time, me and the patient alone, trying to piece together the details of their condition the best we could. A moment of mandatory immersion. I told myself that if there was ever a time I needed to dive in even deeper, it was here.

Many of the patients I saw were victims of trauma though not from car accidents or gunshot wounds. They were traumatized by poverty and their past. Presenting with severe symptoms of anxiety and depression, their pain ached and spread through every part of their life with the ferocity of cancer, which left them hopeless and desiring only to end their suffering by taking their lives. It's called suicidal ideation, and the desires evolve from thoughts to intents to plans. Those who were unsuccessful in completing suicide typically entered through a different part of the hospital.

Other patients presented with psychotic symptoms. One case still echoes in me. He was a man in his late thirties with bipolar disorder. His friends had grown concerned when he became obsessed with starting all sorts of projects. He became angry if any of his friends suggested that he might be spending his money too freely on new business ventures—starting a boating business, for instance, or buying shirts to print and sell, gambling. He started snorting cocaine again. When he was found wandering in the middle of a busy intersection, yelling at cars, the police brought him in. Because the man was combative, we placed him in isolation. I examined him with the junior psychiatry resident. He sat in the room with his arms crossed, hunched forward. The reflected sensations of his eye and head movements were incongruent with what was in the room. He was somewhere else. The junior resident asked what brought him to the hospital, half-anticipating that our conversation was going to lead nowhere. He started an answer and quickly began to erupt with long trains of pressured speech, exclaiming how he had a vision about Earth coming to an end and how he was on a mission to save us all. I tried to get a sense of where he was coming from, focusing intently on all of the mirror-touch sensations, letting all my observations ricochet within me in an effort to bring myself closer into his experience. But it was so erratic. My heart rate increased. My thoughts accelerated as if I were rolling down a hill. Just when I thought I was finally gaining on his high-speed, illogical train of thoughts, the man exploded. He leapt up, the chair falling backward as his leg pushed the table forward. I did not feel threatened; instead, I felt the man's sense of urgency echo through me. Before the resident and I could step out of the room, he stopped. He looked directly in my eyes and shouted, "I have psychotelekinesis like the X-Men, just like you do! Don't think I don't know! I can also see the future and I see you."

He looked at the junior resident, then he tilted his head back at me, smiling. "And you're going to rape her tonight. You're going to rape and kill her!"

He took another step forward, blindly bumping the table again. The resident stuck her head out of the room and called for "Five-and-two, please!"—five milligrams of haloperidol, an antipsychotic, and two milligrams of lorazepam, an antianxiety medication used to sedate someone who is escalating in their agitation quickly and cannot be redirected. Adding fifty milligrams of diphenhydramine as a sedative antihistamine makes the call a "five-two-and-fifty."

I stepped and pivoted out of the room and leaned against a wall as security and an orderly rushed into the room, followed by a nurse. I heard furniture being pushed and a metal chair screeching along linoleum. The team was tying the man down with locked leather restraints. My heart continued to pound wildly. With each heartbeat, my vision bobbed and blurred, and my mind reeled like a projector that had just run out of film. For a few seconds, I lost my own sense of reality and latched on to his and, yes, he could read the future and, yes, I was going to rape the junior resident because this man could predict the future. It was inevitable. The resident probably saw that the color had left my face. "Maybe he's not the best learning-patient for a medical student." I was still trying to piece together what had just happened. "How about . . ." she peered down at her list on a folded piece of paper, "How about you instead go see the patient in room two while I work on this admission." I nodded, probably more times than I needed to. I gathered a blank sheet of paper from a nearby printer—partly to take notes during my interview, but mostly to give myself time to slow down. I stared at the blank page for as long as I could. I needed to come back into my own reality, innocent and untouched. I shook my head, took a deep breath, and stomped my right foot on the ground, snapping myself out of what felt like a trance.

The door to room two was unlocked. I pushed the door open slowly and saw my patient, a bearded gentleman sitting in his chair, legs extended. He was disheveled. He wore a stained olive green button-down shirt over a white-turned-beige undershirt. He stared in front of him. He looked like he was chewing. I felt like I was chewing. I sat down. I folded my paper into four squares, focusing on the sensation of my fingers on the coarse sheet, the sound of my fingernail skimming across the crease, a final act of grounding before I dove into a new set of synesthetic experiences. "Hi. What brought you to the hospital today?" He was still looking straight ahead, gazing with intent at the corner of the room. Then he suddenly looked down at his feet. I tilted forward a little bit to gain his attention and cleared my throat. "Hi. What brought you to the hospital today?" I felt the mirrored sensation of my mouth moving slightly. I looked closer at his lips, which were obscured by his beard. He was mumbling, whispering softly. His left eye and shoulder twitched. He dipped his hands into his pockets and scooted up in his chair. He looked at me. "What's your name?" I asked. His eyes darted over my shoulder. My eyes moved with his. He was mumbling more. I couldn't understand what he was saying. All I could make out was a slow rasping, gurgling under his breath, a voice crafted from years of a rough life, years of yelling and screaming at no one but himself. He twitched and turned toward the left, scratched his left arm, and then settled down. I was there with him, focusing intensely on the shared experience. We both dove into the noise of his internal world. He dodged something that flew at him. There was nothing there, but something had definitely flown at him in his world. There was someone in front of him now insulting him. I felt his face contort in sudden anger, "Now you stop right there!" He put his right palm to ward them off. He swatted at something crawling up his left arm. He put his fingers deep into his scalp and started scratching violently. I felt the pressure of his wild, active internal world pushing up against the edge of his external world, trying to break free. He could not sort out what was internal or external. His reality was ruthless and unwelcoming.

"Fuck you! Fuck that shit no right! You stop lookin' at me!" I looked at him. He was yelling at the wall to his left. He was in the hardest struggle of his life. I saw myself clearly in him and also saw others in him. He was doing the best that he could. I did the best that I could and observed as much as possible to put together a coherent picture of an incoherent multisensory collage. At some point I stepped out of the room, leaving the door open, and took a moment just beyond the doorway. My heart was pounding yet again and my mouth was dry, though not from the agitation of my patient—it was the agitation of my own heart. I found that the thoughts I was used to keeping together and neatly organized had collapsed into one another, amplifying their interlinked emotions. My emotions felt as though they were flipping from one to the next. This was more than the emotional entanglement between a patient and a caregiver, what Sigmund Freud coined as *countertransference.* I wondered again if maybe I indeed had a psychiatric disorder that needed to be diagnosed and treated. Or maybe this was just a further manifestation of the personality disorder I had considered was a possibility: schizoaffective disorder or perhaps features of an avoidant or dependent personality disorder. My thoughts were moving quickly, and I wondered if this was the flight of ideas that I had learned about. And then, just as quickly, I dismissed it as a manifestation of the touted "medical student syndrome." Being with these patients I felt that I was falling out of my own skin into theirs, lost. I knew there was something positive here, that I could maybe treat them from a place of boundless empathy. But at what point would I no longer be able to separate, to reclaim my own sense of self? The onyx sheet in my chest let out a roaring screech, buckling in again. I suffered from their suffering.

I heard my name. "Joel." Was it internal or external? I looked around. Several feet away in the workstation, the junior resident, who had finished presenting her case, was talking to the attending. How much time had passed since I was in room two? I continued to look over at the resident and the attending. They were chatting casually. I reflected the slowed calm in their movements, a smooth fluidity, clean and deliberate. They were communicating with each other with logic that flowed in a natural pace, one piece of information fit clearly with the next, never sliding away, one after the other. I drew my attention more intently on the mirrored sensation of their bodies. I began to settle. I then felt my chest rise out of sync with the attending's chest in an inhale and the warm air of my breath flow out my nose. Yes, this was my body. This is me. I regained the self-awareness to reflect on my own physical sensations, to parse through my internal experiences, no longer relying on other people's bodies.

Focusing away my mirrored-touch experiences on more stabilizing persons was helpful, just as effective as staring at a single point or stable object. Thinking about my body, my own physical sensations, was also useful and grounding. Though I realized I still had a long way to go. During the rest of my rotation, I was able to get through by remaining acutely aware of this process. If I caught myself slipping, I practiced intense visualization, such as imagining myself enshrouded in a thick velvet cloak or wrapped in a dense cocoon, preserving myself from the external world and the leaking internal world of others. I insulated myself in my own experience, my own reality, which allowed me to follow a conversation with patients long enough to guide a structured conversation to uncover what the patient might have otherwise obscured: what was actually going on at home, what unarticulated symptoms they were afraid to confront, where their greatest source of joy was located, where their greatest source of pain lurked.

Einfühlung, a German term coined by the son of philosopher

Friedrich Theodor Vischer, speaks to our ability to generate meaning by infusing our own thoughts and feelings into others, including objects and art. Later translated into English as empathy, *Einfühlung* refers to a mutual exchange between the observer and the observed. Because meaning never arises from the external world, we receive information from the external world and create a meaning in ourselves, which we then associate with what we were observing. Friedrich Nietzsche alluded to the case of excessive immersion through *Einfühlung* as spectators merging into whatever they were observing until they exhibited a "loss of speech and the dissolution of individual identity." Described by Nietzsche, this kind of *Einfühlung* is a positive experience—the identification of the self within another served as proof of the importance of the other person and of the relationship between the two. The more intense form of *Einfühlung* was often described as going beyond "seeing" and actively "looking" to encompass an experience of physicality that allowed the observer to penetrate the world of the other person. A communion of similarity established between the self and the other in concert with the unity of physicality and imagination opened up this point of entry between two people. In essence, looking (or attending) provides the sensory information that then allows not only for empathy to exist but also for imagination, the cognitive processing of the mind, to take hold of this information to create a mutual relationship with the other and imbue the experience and the other with heightened emotional, personal value. This relationship between the senses and physical experience of these senses creates a sort of self-generated doorway in one's own mind that leads into the body of another as long as the experience is vivid enough, regardless of whether the re-created simulation is accurate.

Empathy (and, by association, compassion and kindness) requires boundaries. Rather than serving as separating walls or dividing lines between yourself and others, boundaries are actually more like a necessary skin between yourself and the surface. Boundaries are the clearest demarcation possible between what is comfortable and what is uncomfortable—what violates your mental space, including your morals and values, and what violates your physical space, especially your body. The more blurred professional boundaries are, the more likely we are to fatigue our capacity for empathy. Much research has been conducted in careers that require empathy concerning the subject of *empathic burnout* or *empathic fatigue,* even to the degree of causing secondary trauma. The more aware you are of your own mental and physical experience, the more familiar you are with your boundaries, which ensures the integrity of your own self.

This is a difficult task for anyone, though it is admittedly more complicated for people with mirror-touch synesthesia. In my case, seeing another person hurt hurts me. Seeing a person suffering, I experience pain—especially watching people who lack or obscure their own voice, either psychologically or symbolically. Like a container with holes, their emotions pour through me from all sides, as if the experience and pain were my own.

My rotation through psychiatry offered me an invaluable lesson in understanding the importance of boundaries. At the same time I was so humbled by the power of my overwhelming and prolonged experiences of heightened empathic states. Though regularly unnerved by them, I couldn't help but marvel at these experiences. To approach another with empathy is a wonder, an essential part of who we are as humans. But, just as oxygen sustains us, too much can kill. I strengthened my resolve to continue practicing empathy. But I was honest enough with myself to know that I needed to apply it from an angle natural to me, that I could sustain and respect.

On the last day of my psychiatry clerkship, I picked up a small piece of blank paper. I folded it to the size of a business card and drew pictures to represent what I saw as my goals. There was a brain and there was an arrow pointing upward—visual symbols loaded with meaning that I knew would continue to keep me focused in my pursuit, to keep me from derailing myself, to remain loyal to the internal voice that helps me figure out my next best step. I tucked the piece of paper into my wallet where, if I ever needed to be reminded, I could look upon it.

Admittedly, I did walk away from the rotation with a little bit of residual discomfort. I was shaken by the realization of the immensity of how much more I still needed to learn about my trait, which I couldn't find in a medical school textbook. I needed to be even more curious about myself and others before returning into the world of psychiatric illness. And yet, I was also more confused than ever about whether my synesthetic experiences were an ability or a disorder, a condition to be treated, a blessing or a curse.

CHAPTER 6

A Blessing or a Curse?

I SAT IN THE MAIN AUDITORIUM OF TATE MODERN IN LONDON surrounded by artists, musicians, professors, researchers, and a smattering of tourists curious about modern art. We all stared up at the stage, which was covered in red carpeting that hummed vibrantly under bright spotlights, waiting for the interview to begin, one of several scheduled that week during a symposium on mirror-touch synesthesia and the trait's relationship with art.

James Wannerton, the longstanding president of the UK Synaesthesia Association, opened up a leather folio and scanned his list of prepared questions. A *lexical-gustatory* synesthete, James associates written words and letters with distinct flavors, sometimes as specific and powerful as malt vinegar, warm semolina, hair spray. His guest, a young woman named Fiona, was dressed in a black leather jacket, adorned in boots that ran halfway up her calves. She crossed her right leg over her left in anticipation of his first question.

I sat in the second row, captivated by Fiona's humble plum 3 arranged with radial soft friendly blue 4s, which were speckled at the fringes with turquoise 7s, an eccentric and endearing combination. Fiona was the first mirror-touch synesthete I had seen in person. I felt as if I were meeting a fourth cousin once removed, a distant member of my clan. A riffle of familiarity and distance stirred in my stomach. My fingertips vibrated, my toes tickled. I felt the sensation of Fiona's eyeglasses sitting on the bridge of my nose. The light brushing of her hair swept across my forehead followed by a soft pawing and patting of her outer thighs, which betrayed in my own a slight glimmer of stage fright.

When she finally spoke, a thick stillness slipped over the audience, and I heard a familiar chord in a foreign melody.

"This is what it is. Or, at least what it was," she explained. "I had this experience in America. I was sitting in a car and there was this fury going on between two men over a parking space. One man punched the other, and that was it. I felt it. I felt punched. I passed out. All I had seen was the punch. That was it. The medical team couldn't place it. They thought I had perhaps had a seizure of some sort. It was only when I returned to the UK that I learned of my mirror-touch synesthesia. I felt validated."

I looked across audience members and felt the flickering of their wonder at Fiona's mirror touch, wide-spread wonder and an unmistakable pity. Meanwhile, I wondered if I had been blinding myself all along. Was mirror touch a *condition* to be pitied? A disease to be treated? The thought left an acrid aftertaste in my mouth. Maybe people were right—perhaps mirror touch was not so much a disease as a tragic hex.

Some synesthetes feel strongly that their synesthesia is absolutely a blessing. Perhaps they found a way to relieve the tension between their internal and external worlds, bridging the neural chasm between "real" objects and their synesthetic qualities. Graham Harman, a key philosophical figure in the development of object-oriented ontology and speculative realism, thinks about the neural chasm in metaphysical terms. He claims that humans are incapable of experiencing the "realness" of an object in the physical world. We can only experience an object's sensed qualities, their "sensuality." For example, how prickly a cactus is, or how blue a sky appears overhead. We can, however, experience the opposite. We can experience *sensual* objects that have *real* qualities whether or not we believe they exist as *real* objects, such as imagining a basketball in our hands and experiencing its color and weight or imagining a unicorn and experiencing its touch and smell. The basketball is considered a real object. But as mentally sensed objects, both the unicorn and the basketball have *real* qualities. Similarly, synesthetic sensations exist with *real* qualities. But synesthetic sensations can only come into perceived existence when the brain assembles together their qualities through hyperconnected synesthetic neural networks. In other words, synesthetic sensations are only perceived if an entangled synesthetic brain response is triggered in full and crosses the threshold into conscious awareness. Without triggering these networks, the synesthetic sensation registers with little or no effect, slowly fading from our minds as harmlessly as idle thoughts slipping into irrelevance. I feel the physical sensations of other people's movements, their clothes, their postures. But while I might experience the sensation of someone picking spinach out of his or her teeth, as if phantom fingers were messing around in my own mouth, the feeling of a person's shirt collar might not resonate within me. And yet, when synesthetic sensations trigger these hyperentangled neural networks with force, it is often impossible to determine the difference between what is real and what is, arguably, not real. To successfully differentiate between the two, synesthetes require the smooth, organized hum of background mental processes. And when such experiences are particularly vivid, synesthetes require an even more commendable feat of conscious mental mastery, particularly when synesthetes like Fiona, in order to gain new insights about the world or themselves, immerse themselves completely in a synesthetic experience.

Fiona went on to describe how she avoided cities and crowds throughout most of her early adulthood. Though here she was, casual and at ease in a room full of squirming, wriggling strangers, their bodies pulsing all around her. The longer she spoke, the less I felt the fidgeting sensations on my thighs or hair brushing against my forehead. They were slowly replaced with more fluid movements. Fiona was all smiles and hearty laughs, sharing the quirks in her daily routine, punctuated by moments of earnest gratitude about her mirror-touch experiences, which she told the audience have allowed her to get closer to her friends and family, to better understand them emotionally as people because of her trait. My shoulders lowered, and I was now nestling into my chair. I relished in the generous doling of smiles and giggles resonating through my body. The warmth and coolness I so enjoyed bubbled upwards in me like a sparkling *moscato*. I was impressed at how brightly Fiona allowed her 4 to shimmer. It seemed as though she had somehow managed to hold close to the blessings of her synesthesia while casting off its many prickly thorns.

Studying her ability to speak so openly, so personally, in such a public space, I felt a twinge of envy. I was scheduled to speak later that day. The head of the symposium was a talented artist who invited me to share my perspective, but I was unsure of what she meant. Seeing Fiona, I knew that when it was my turn to speak I would still be muffling the voice of my messy, internal, synesthetic world. My default was to extend my "professional warmth," to approach public audiences with contact precautions, thick latex gloves, elbow-length if possible. But Fiona, her 4s danced elegantly in cerulean stardust. They gleamed with a cool luminance, glittering in a brilliantly harmonized chorus as if chanting, "Don't let yourself be erased." I was transfixed by her 4s. If only I could be like them, I thought, an arrogance of selfhood—a lightness of being. How did Fiona reach such a seamless level of comfort with her mirror-touch synesthesia, with speaking about it out loud, and in public?

At the end of the symposium, organizers introduced me to Fiona. Because neither of us had ever met another mirror-touch synesthete, they seemed giddy to make the connection. We shook hands cordially and said our hellos. Lingering near the stage, we talked awkwardly, staring at one another, incredulously. Averting prolonged eye contact, leaning slightly away, pivoting our shoulders and feet toward the exits—I felt my own shyness mirror on her and reflect back on me. Or was it her shyness reflecting on me and mirroring back at her? Where did the two mirror-touch synesthetes begin? Before we had the opportunity to move beyond pleasantries, we found ourselves split in conversations with the other attendees. All told, Fiona and I only got to spend a few anticlimactic moments together, most of which were spent probing each other to make sure we were the same species—clumsily riffling for insights about ourselves through the other. Before we left, though, we expressed a shared desire to see each other once again, to eschew the pleasantries next time and ask the incisive and personal questions that we have only ever quietly asked our own reflections.

And yet, I couldn't hide my disappointment, its cold hangover. I had wanted to learn how to reconcile the difference between the real and the not-real while still being able to experience the full realness of my synesthesia. Mental mastery seemed attainable, but it was deceptively elusive. I had been naïve, perhaps even infantile, to expect so much from a single person, to have someone offer up to my lips a cup filled with the insights I was looking for, distilled.

My trip to London was designed to help me get a better handle on my trait. I worked closely with noted mirror-touch researchers Jamie Ward and Michael Banissey in their lab at University College London. I was still struggling in the hospital, trying to learn how to extricate myself from my synesthetic tangle with the patients, with coworkers. I wanted to be able to dive in and out of these experiences freely without any lingering synesthetic sensations ricocheting throughout me for the rest of the day.

I decided to take up Ward and Banissy's invitation to participate in their research, just as I had done in Ramachandran's lab a few years earlier. For a dedicated week I performed lab exercises and experiments, from low-tech tasks that measured my introspective ability to estimate my own heart rate to elaborate illusions with curtains and false hands that tested my sense of agency and ownership of my own body. During the hand-illusion task, I sat motionless in front of a full-length mirror with a powder-blue sheet draped over my body. Only my head was exposed. Behind me hung a similarly colored curtain with two holes cut out of it. The holes were located right around my shoulder level. The experimenter, wearing a powder-blue blouse with white gloves sewn on, placed her arms through the holes, bestowing on me a pair of illusionary hands. Looking at the mirror, I listened to instructions through a pair of headphones slung over my ears. "Make a waving gesture," the voice said. My illusory hands waved at the mirror. "Point to the mirror." This time, my illusory hands pointed at the mirror. "Snap your fingers twice." Instead of performing this task, however, my illusory hands made a beckoning gesture. This exercise continued for many rounds. Sometimes the illusory hands performed the correct action; sometimes they performed a different action than what the speaker demanded. I was asked a structured series of questions between rounds. As the experiment progressed, I felt more in control over the illusory hands, particularly when the action matched the instructions. But at the same time, even when the action and the instruction didn't match, I always felt as if the illusory hands were my own. The findings of this psychometric experiment supported the general hypothesis (as well as my own personal hypothesis) that mirror-touch synesthetes likely experience a breakdown in brain processes responsible for differentiating between the self and the other, resulting in a blurring of innate boundaries.

During another experiment, I performed the infamous visuo-tactile Stroop task. This task, which measures differences in reaction times and accuracy on a match/mismatch task, can objectively differentiate mirror-touch synesthetes from non-mirror-touch synesthetes. With "tappers," or small pins in plastic attachments taped to my face, I watched recorded images of a woman getting touched on her left or right cheek, sometimes both. Throughout the video, the tappers—true to their name—tapped my left or right cheek, sometimes both, while I observed the woman getting touched. After each tap, I had to press a button to indicate whether the *real* physical tap on my face occurred on my left cheek, my right cheek, or both. Sometimes the touch on the woman's face mirrored exactly how I was tapped. This felt natural, seamless. Her left, my right. Her right, my left. Other times, the touch on her face might randomly occur on the same anatomical side as the tap. Her left, my left. Her right, my right. After each tap, I immediately pressed the button, trial after trial, congruent or incongruent with the woman's touch. The permutations were dizzying. The taps and images occurred far faster than I could process, let alone be consciously aware of what I had physically experienced before I had to press the button. Before long, I could no longer differentiate between real touch and mirrored touch. The results were validating. "Your results clearly indicate that you are a mirror-touch synesthete," the graduate student performing the tests told me, "both in terms of errors and in reaction times." Though I still didn't know if my mirror touch was a blessing or a curse, I could at least finally confirm it was not a delusion.

As reassuring as this was, I was struggling with my trait in my personal life. Just as I need to understand my mirror touch to survive my professional life, I needed to recognize more clearly the difference between me and Jordan. When I was in London, he and I had been married for a few months. Our wedding was meant to represent our solemn and joyous vow to build our lives together. Though leading up to our wedding I had doubts, which grew and multiplied over the early days of our marriage. I was sinking in uncertainty, yet I continued to convince myself I wasn't, just as I had done with Cristina. By then I had become a master at distorting my reality with the same unconscious efficiency I had developed to identify synesthetic sensations from physical sensations. Yet even in my own distorted reality, I could feel a suffocating pressure around me as I sank deeper into my relationship, into Jordan. Relationships require work, and this work was mine: to love myself first and still show Jordan unconditional love. This meant I would have to help him be more of himself while still helping me be more of myself. But, how to reduce the effects of my mirroring reflex?

Like praying a rosary to ward off the inevitable, I told myself over and over how I had learned from my relationship with Cristina, how I knew better than to lose myself again in another person. Again and again, to prevent my own incarceration through entanglement, I refused to become one with him. But it was already two years too late. To learn more about Jordan, to fully understand him as a person, I let myself experience the mirror-touch sensations to their full extent. To be a better partner, to be a better husband, to love him with the same love I gave myself, I had to see myself in him. I assured myself this was different than my other experiences, not at all like the ones I experienced with Cristina, not at all like the ones where I nearly lost myself in the psych ward.

As the wedding drew closer, I laid rescue mechanisms, triggers to pull me back to the surface if I risked losing myself completely. I had to outsmart myself and my synesthetic tendrils. I set rules for myself, proclaiming that if he violated any of those rules, then the wedding would definitely be off, then I would absolutely leave. Within hours of our wedding vows, however, I found myself muffling my own weak cry for help with the debris of shame and an unconditional absolution that were simultaneously mine and not mine. I lost track of where the pieces came from.

As much as I sought answers from Fiona, and Banissy and Ward, I left London with even more questions. I wondered how long I could continue to drift between the real and unreal—lured deeper by faint glints of synesthetic insight while drawn further away from a distant, glassy sky.

A year later Fiona and I finally met again, albeit through a computer screen, which, as it turns out, is the ideal way for two mirror-touch synesthetes to interact. Fiona sat comfortably in front of her screen, backlit by a floor lamp. Even through the computer screen, I could still make out her facial features, feeling the echo of her movements and expressions on my own face more than six thousand miles away. I could make out in the far corner of her room a shelf with books intermingled with neatly placed Harlequin dolls. An opaque beige curtain with orange stripes blotted out the light from the street outside. We began the conversation by going through our days. Fiona told me about her thriving pet grooming business, which she operated from home when she wasn't preparing the upcoming defense of her thesis on the intersection between learning disabilities and workplace technology.

To focus on what she was saying, I stared at the power button of my computer monitor, which glowed alabaster. When I looked up briefly at the screen, I realized she was doing the same thing—two mirror-touch synesthetes at their best, communing without eye contact. Suddenly aware of one another, Fiona and I sat up in our chairs, laughed at our behavior, then settled in for our much-anticipated conversation.

Fiona had scrawled a list of questions in a composition notebook. But I only had one immediate question in mind, a lingering query from my time in London the previous year: How had she managed to turn her mirror touch from a "curse" to a "gift"?

"I've had to teach myself quite a bit to get by with my mirror touch all the way from a young age," she told me. "I can still remember quite clearly one of my first mirror-touch experiences. Television first came to South Africa in the late seventies and my family got one. I remember we were watching a film, *Ring of Bright Water,* and that was the first time it stood out that I really couldn't watch or see certain things. In the film, the farmer was busy using a spade and cultivating soil, and there was an otter. He used his spade to kill the otter. I remember I was horrified. I *felt* the farmer killing the otter. It felt as if *I* had killed the otter. My mom couldn't console me. It went on for about a month. She would tell me, 'Fiona, remember the three Fs: feelings are fickle, Fiona.' She would say it oft to me. It was only a film, but it felt very real. It was the first time where seeing something really affected me. And, you know, even though I could feel the experience of others, I felt different from others, too."

Fiona's early experiences with synesthesia initially made it nearly impossible to inhabit the "real" world.

"With sounds, cars going up and down the street, it feels like something's rubbing against me when they pass, rubbing against my shoulder blades and arm. If cars pass the other way, I feel the rubbing coming up the other way. I watch somebody when they're walking and their legs are moving and they're making sounds with their footsteps. It's a very odd one actually. It feels like things are moving on my body, you know, touching and moving. With tastes, as a child my mother used to give me incredible loads of sugar, sometimes sugar on food and bread and all sorts. I make myself very strong sweet tea, two teabags and about three sugars, and when I taste the sweetness I have a melting feeling. It's more of a melting feeling like a soft ice cream. I don't experience it as being specifically cold, rather, really soothing softness. Whereas I can go in a supermarket, and I've got to whiz by the meat section because even though it's packaged, I feel as though all the raw meat and its smell is stuffed inside my nostrils. I'll walk past aisles of food and look at it, or even at restaurants looking at people eating, and all of a sudden have a taste on my tongue that doesn't quite fit or the sensation that I'm eating when I'm not. There was never really a way to turn that off."

Fiona's experience reminded me of CC, another polysynesthete. CC does not clearly have mirror-touch synesthesia, but she does have *synesthesia-for-pain,* a unique type of synesthesia, in addition to other forms. When she sees another person wounded or in pain, she experiences predictable tactile synesthetic sensations in her body, though—unlike me—not in the mirrored locations of the other person. CC remarked on her noisy synesthetic world with a slightly different slant than Fiona. "The thing about my synesthetic perception is that, maybe because I learned how to read early, my sense of numbers and letters for me were always really brightly colored, though the colors themselves are not necessarily the most vibrant or elemental. For example, I don't have a single letter in my alphabet that is a true bright primary red. I don't have a single color in my alphabet that is a true primary yellow. Regardless, the sense of colored letters and numbers was there from my earliest associations, and so I feel like things have always been so easy to remember, because there is something else there and not just simple black ink on a white page. I think my memory, that exceptional memory that I have, is from my synesthesia. I don't know if it is about anything else. If you think of memory as being a row of hooks, we all get to hang things on these hooks so that they are there for us when we come back to it. Well, I think people with synesthesia have more hooks—more hooks to hang things."

I considered Fiona and CC's different childhood experiences. It seemed as though grasping what gifts could be gained from synesthesia was far more natural for CC. Maybe some of us are simply built from birth to be able to manage synesthetic curses better than others. Or perhaps, by the grace of good fortune, some synesthetes are bestowed with only the gifts, a form of synesthesia that goes down easy—a seedless, boneless, skinless synesthesia. My grapheme-color synesthesia, for instance, has been fairly harmless since childhood, even if I do still prefer to color my letters in the colors that match most closely my synesthetic colors.

As Fiona and I continued to talk, her camera suddenly went black. A large fluffy cat with raven fur sashayed into view. It turned and wafted its tail at the camera. Fiona scooped the cat off the desk with maternal finesse. I felt the sensation of weight on my right shoulder, followed by the hysteria-inducing tickle of fur under my nose as the cat climbed up Fiona's left shoulder, wrapped its tail up under her nose, scooped up its hind legs, and pounced downward onto the ground.

My nose still tingling, I asked, "How does your mirror-touch experience with animals compare to your mirror-touch experience with other people?"

"With people it's stronger, but with animals . . . When I say it's stronger with people, there is a distinct difference. It's the 'nonverbal.' People will talk and their words don't always coincide with their nonverbal communication. Animals will wag their tails. Some will bark, some may let out sounds but, even if it's mainly nonverbal, their emotions are easier to interpret because they match what I see. So, I'm actually far more at ease with animals than I am with people."

Fiona trailed off for a moment.

"One of the things that kids discovered was that if they harmed an animal in front of me, it would cause me to have a severe reaction. They used to find ways to make me squirm in pain of all sorts. It was horrendous. One would chop off a lizard's tail in front of me. And then there was my rat."

I felt a phantom swallow, though my throat remained steady. I felt and saw her eyes drift to a corner of the room. I could feel the distress coalescing from deep behind my belly.

"I had a rat. She was white, a female with pink eyes and a long tail. She went everywhere with me. She was always on my neck, you know, sitting there. I would have these little talks with her, and she always would just sit around my neck. One day these boys came up from behind and snatched her off my shoulders."

As Fiona relayed her story, I felt our facial expressions detach from her words, incongruent. The tone of her voice took on a more measured tone, flattening out into two dimensions. "They burst . . . they basically burst her body."

"*Burst* her body?" I asked in horror.

"They squeezed her right in front of me. Squeezed her until she burst. And that was horrible. Not only was it painful because that was my pet, my friend, but it made me feel like my entire body burst as well. I couldn't breathe. They held her right in front of me. One of them was holding me so I couldn't move."

"How did you ever recover from that?"

"I developed a lot of anger. For a lot of reasons, anger at adults, anger at people older than me. Because I seemed so sensitive, my father tried to toughen me up and harden me, which made it even worse."

Fiona crossed her legs up onto the chair, shifting until she found a comfortable position.

"I often wonder, when you've got somebody with something that makes them as sensitive as what I am, or you are, and they're exposed to conditioning, whether that leads to having no empathy at all or having empathy for everything. I find that, no matter what I've been through, empathy prevails. It's there all the time. It seemed more effective than my father trying to toughen me up. It not only helped me deal with some of the emotions like the anger but also to understand them. I think this is what the empathy is about. If you can understand what makes a bully a bully, what makes somebody behave that way, what makes them target you. It's not that you sympathize with the person. It's just that you can understand how it can arise. It's not even about forgiveness at that point. It's just a part of living, people and nature."

The first time Fiona became aware of the importance of empathy in her life was in her childhood. Her father would come home from work, exasperated, demanding two things: whiskey and to be left alone. While pouring his whiskey, Fiona's mother would vent about her day, her frustrations of being at home with two small children, 'Fiona did this and this today. She's been really bad.' And Fiona would get belted. Fiona considered that maybe he was just trying to purge his anger out by belting her. But perhaps it was more than that. Perhaps this was all he knew.

"I could deal with my father hitting me. In my mind, that is. I understood why it was happening. I understood what would lead up to it. I would have a sort of monologue for myself. 'Mommy's going to make daddy angry again because she can't help it. It's the way she is. Daddy's going to hit me and it's going to be okay.'

Don't get me wrong, I was still very frightened of being hit. I would wet myself. I would be frightened because it would be so unexpected. I would talk myself through a lot of these situations. It was a way of dealing with it. However, it doesn't mean that if you've had trauma in life that you don't need to speak to somebody else. You need to get the valuable perceptions of other people. Any of us can be perceptually blind. Just because I've got mirror touch doesn't mean I've not got my faults and whatnot."

Fiona felt that synesthesia helped her through that time in her life. She would feel her anger, like an ocean current. She felt like she could swim through it or ride it like a wave. The current transported her elsewhere when she was being hit. At one point, after the beating was over, her father became enraged because he could not get a reaction from Fiona. She explained, "The reason he got no reaction is because, well, I think there is pain and then there is *pain*. When he belted me, I felt that his pain was far greater than my pain."

Through her mirror-touch sensations, Fiona could appreciate that the pain of her father was far greater than the pain she felt on her skin.

I thought about Jordan. By this point in time, well over a year had passed since our wedding day. I found myself frequently drawing strength from empathy, though with each transaction the merging continued. I had matured to love myself, to be forgiving of myself, to show compassion for myself. As a result, I felt that same love and forgiveness and compassion for Jordan. The more love and tenderness I gave him, the more love and care I felt I was giving myself. But then, where was this pressure coming from? Was it coming from the edge of myself? Where had the boundary gone? There was a subtle itch embedded in each of the mirrored echoes of my hands on his body, mouthing the words, "What about me?"

"When I reached the age of eleven," Fiona continued, "I had to put an end to the belting. I had a chat with him. I remember him saying to me, 'I didn't realize you thought that way. I didn't realize you could think at that level.' It was a strange realization for him. It was then that we finally started having deeper conversations. We became a lot closer. Perhaps I was a lot older than he thought."

I felt the smile on Fiona's face, though her smile was betrayed by the fatigue in her eyes. To change the subject, I asked Fiona if she still spoke Afrikaans.

"I do and can understand Dutch words. In the end, it doesn't matter as much, because when people are expressing themselves I really only experience the emotion in their faces. I don't know if you've ever sat and watched people speaking a foreign language. But I think because of the mirror touch, I can read their body language as clear as I can read."

I was relieved that Fiona and I had this feeling in common—the mirror touch acting almost as a solvent, dissolving boundaries of language, exposing the raw nonverbal language shared between people. I recalled a Russian-speaking patient I once saw in clinic. She was an older woman who had suffered a small stroke but, as in much of neurology, size is not nearly as important as location. Her stroke was in the cerebellum, the part of the brain responsible for balance, coordination, and sense of equilibrium, which is tucked below the visual regions and next to the brainstem. She was left with a constant feeling of dizziness. We communicated with the help of a Russian interpreter. She exclaimed with joy at how she had been exercising and watching her diet to prevent the potential for a future stroke. Because we were already running close to the end of the appointment, I could have checked another box and moved on with documenting the encounter, but something was off. There was an abrupt change in my mirrored sensations—from elevated shoulders, eyebrows, and smile to a still face and shoulders melting into a hunched position. Without speaking Russian, I was concerned about what she was likely saying behind the seemingly impenetrable wall of language. "I'm so glad you've been able to stick to your diet!" I congratulated her. I paused for the interpreter to translate. "But, I know what you went through recently and are still going through must have been really hard. Aside from the dizziness, how are you feeling?" I paused again for the interpreter, though just as the interpreter had begun to translate, she broke down into tears. She wailed in Russian as the interpreter translated that the only reason she was able to stick to a healthier diet was because she was being pinned down by the fear that she was going to have another stroke. She was in pain, a deep agonizing pain that flooded her with the terror of her own death.

"There are times, though," Fiona continued, "when my synesthesia probably misled me. When people unsettle me, I get this distinct feeling that there is something unsaid. It makes me feel uncomfortable. But I'm never sure whether it's true, or what it is, really. I think it's very hard because you've got to always think about what you're reading and where it's coming from, especially if you're doubting somebody or a situation."

Yielding to my rational side, I was doubly relieved that she also shared my trepidation about blindly trusting the accuracy of this impression. By the time I was in high school, I understood the consequences of speculating about someone's internal mental state. I recall annoying a few of the more self-protective guys who prided themselves on their masculinity, the illusion of imperviousness, of invincibility. Even if in the name of healing or friendship or curiosity, to make a gross error could have terrible consequences, like being derided as a "faggot" or, later, shoved and rammed against lockers with their unforgiving locks and latches. I had to be selective about when to say something about my synesthetic reflections. I needed to trust in the neural networks underlining my intuition while always preserving the humility to verify from a place of kindness.

Fiona was not all that different. She drew parallels between reading people and reading in general. When looking at someone, she might pick up on movements or expressions that are out of place. She grasped that people's movements changed under different emotional conditions even if the change was at a level that most people would not be aware of. There were times that she would take the risk and ask friends if they were in distress. They might disabuse her. But within a few days, they usually admitted she had been right. "I find it very hard to *know,* really. I actually do trust my instincts. I trust it now more than ever, but it's a hard one because sometimes you've got no evidence. There's no particular grounding for having this sort of synesthetic experience that might tip you off. Still, I tend to evaluate things and try not to take it personally if I'm wrong."

Fiona naturally grasped the principal behind Ramsey Theory. Given enough information, the theory posits, there is almost a 100 percent chance that a compelling pattern will emerge. To survive, humans evolved the capacity to notice patterns amid a flurry of noise. Our brain is wired to discover intention or significance even if there is none. "I find it's better not to say anything most times and let it unfold," Fiona described, "though I remain aware and may alter my interactions to people." She might purposely alter the way she communicates. She feels that simply being aware of others' vulnerabilities helps to cushion the impact when they become rude or abrasive. "If anything, it's de-escalating their response. Without my synesthesia, I don't think I would be able to do that."

I was touched and grateful by how open Fiona was in sharing these meaningful details of her life. As I listened intently, I found myself placing tokens in both weighing dishes of the mental scale balancing between gift and curse. The scale teetered delicately, though I wondered if others would have stacked as many tokens as I did onto the "gift" side.

Still, because synesthesia is partially—if not unconsciously—bidirectional, the sensory noise can be inadvertently amplified even further into what can feel like the clawing, blinding screech of audio feedback. Words give rise to tastes. Tastes give rise to words, which then give rise to tastes once again, round and round in an unbridled ouroboros of associations. Each experience can become more distressing if the associations are uncomfortable or repulsive to the synesthete. CC's synesthesia-for-pain, for example, quashed her childhood dream of becoming a doctor. If CC so much as saw a cut on a person's hand, she would instantly feel a bolt of electricity run down her spine, down through her legs, a rapid and extremely painful zing or zap that wrapped around her like a shock. The gorier the sight, the greater the reaction. "It was horrible," she told me. "I had a brain that was on fire."

At the mercy of internal associations, synesthetes tend to share an experience of feeling alone early in life. For me the loneliness seemed to stem from the perception that few if any people around me shared perceptions even remotely similar to my own. It was often difficult to understand other people. With the added layer of mirrored sensations, mirror-touch synesthetes encounter an additional interpersonal paradox. Throughout my childhood and much of my early adolescence, I found myself stranded at the fringe, tugged between feeling very different and very much the same. I was unique, but also incomprehensible; there, but untouchable, inherently unknowable. I avoided putting myself into situations where most people connect with others. Exposing myself to the potential of so many unpleasant feelings was like willfully feigning rest while lying atop a bed of hot nails. The anticipation of pain is often worse than the pain itself. I wasn't naïve. I knew I would survive, that my life was not actually in danger. But it hurt. Pain ignites fear before lighting the fuse of hatred and anger. Pain reflected from others complicates the pain of isolation and the pain of rejection. It was enervating. Withdrawal was tempting. Feeling the pain of another on top of my own pain reliably triggered my most primitive reflexes to avoid pain, which meant I had to avoid others. The greater the pain, the greater my desperation to escape. We can perceive this desperation as being "aloof," "snobbish," or "just not feeling well." But behind the door with a porthole, covered and sealed shut with plastered labels, is the sound of fingernails feverishly scraping wooden floorboards—our brain's singular primal motivation to escape all pain.

When younger, I assumed that I was defective or deficient and, to use CC's metaphor, somehow not flame retardant enough while others seemed to enjoy leaping into the fire. Amid the isolation, I sought solace by satisfying a longing to make the external world match my internal world. Coloring letters the "right" color, deliberately mirroring the "right" bodily positions of the people I see or the sounds I hear, physically touching the parts of my body covered in phantom fingerprints, anything to relieve the tension. In retrospect, psychoanalysis offers a framework for thinking about the relationship between synesthetes and their synesthetic perceptions. An acceptable behavior, value, feeling, or perception is said to be *egosyntonic,* a harmonization with one's self-image. Synesthetes are considered egosyntonic when they feel that their synesthesia is an essential part of who they are. The challenge then arises from its opposite, *egodystonic,* when values, thoughts, feelings, perceptions are in conflict, or dissonant, with one's self-image. For example, egodystonic reactions are common in persons with Tourette's syndrome and who may be distressed by their tics. It is also found in people with *misophonia,* a repulsion toward specific sounds (such as chewing or the long "e" sound in words) that is severe enough that it impairs daily functioning.

Before reconnecting with Fiona, I met a young mirror-touch synesthete named Abishai. Like me, a polysynesthete, but unlike me, still in the throes of adolescence. Abishai experiences intense feelings of discomfort whenever he's exposed to the high-pitched din that emanates from fluorescent light. His distress from the synesthetic sensation buzzing through his entire body, his desire for it to end, are part of what make the experience egodystonic.

Abishai was able to provide an insightful frontline perspective about common challenges faced by newer generations of synesthetes. For instance, the emotions mirrored on mirror-touch synesthetes have the potential to be incredibly overwhelming. When Abishai was only eleven years old, he found himself sinking into the quagmire of other people's lives. A girl in his class was in a motor vehicle accident that injured her spinal cord, paralyzing her from the waist down. Any time Abishai saw her, he was smothered under the weight of her disdain and sorrow. Yet, she was never outwardly angry or visibly full of hate. To most people, she appeared upbeat, resilient. Around the same time, another classmate was hospitalized for an unknown devastating medical illness that effaced whatever trace of youthful bliss he had. Abishai, in turn, was bombarded with the feelings of now two classmates who were involved in tragic circumstances. Abishai felt as though he had experienced their trauma firsthand. He would ruminate over and over again about how he could help them, how he could make them happier, how he could make these dark feelings ricocheting inside him go away. He became progressively more distracted. He was unable to focus on what his teacher would say in class or what tasks he needed to complete.

The mirror-touch distractions were generously layered on top of his medley of other synesthetic experiences. He would try to start his homework but would inevitably become bound in obtrusive thoughts about the taste of a word or the color of a sound. Math felt impossible. Like me, Abishai has ordinal linguistic personification. If numbers represent colored personalities, entire living and feeling personas, then how does one perform arithmetic? What is 2 + 2? If Abishai's yellow and overly loquacious 2 is added to a similarly loquacious 2, then what you end up with is a chatty, jaundiced couple of 2s, not the number 4, and certainly not a green 4, assertive and politely extroverted with a penchant to teach others.

Outside of school Abishai continued to experience an almost insurmountable volume of synesthetic noise. Anyone he spoke with, he would quickly lose himself in the synesthetic tactile sensations crawling over his entire body like a heaving swarm of termites. For years he would come home from school, hide in the basement and turn off all the lights. The silent darkness rescued him from a relentless undercurrent of senses. Out of desperation he clung to sensory deprivation. "He has this war face," his mother confided in me, "like he's still just getting through the day. It takes an extreme amount of concentration and energy from him."

The subjective experience of the "ego" or the "self" can also be affected through its very dissolution. Mirror-touch synesthetes, myself included, report that when we are completely consumed by synesthetic associations, the boundaries of who we are disintegrate. There is a sense of oneness with our surroundings, a borderless existence, where it is often impossible to tell where our bodies end and the external world begins.

Fiona recalled an instance when she was temporarily living in a remote cottage in the farmlands of Essex. "I know this is going to sound really odd, but when I experience pain I also experience blue or orange. Orange is when it gets really severe, you know.

My collie was still a puppy at the time, but almost overnight there was this thing where every time I looked at her I kept seeing these . . . snakes, oh I know it sounds odd, but I'd look at her and I'd have the synesthetic experience of these two snakes almost colliding, and I kept seeing this bluish-orange around her. Mind you, I didn't see them literally in front of me like a hallucination. They were in my synesthetic mind's eye."

For two days she kept seeing the mental image of two colliding snakes and bluish-orange around the collie. Though she appeared to be eating, drinking, and getting along just fine, Fiona maintained her suspicion that something was wrong. She described the experience of colors as "an almost instinctive feeling." After the second day, the collie became extremely ill. Fiona scooped it up and took her a very long way by taxi on the Sunday before Christmas to another farm where there was a vet. The collie had been poisoned. Fiona discovered that the farm she was on used a pesticide on their land. Once Fiona found out the specific poison, the vet was able to neutralize it. The collie was fine, but Fiona was stunned by her experience. She explained this calmly, matter-of-factly, though with a hint of hesitance, an internal skepticism. "It was the strangest thing. It was *almost* like I had this pre-sense. And I think my synesthesia had everything to do with that."

While the idea of ego-dissolution and precognition can evoke images of psychics and mediums delicately fingering their temples, we can at least speculate that the synergy of Fiona's synesthesia, and perhaps a natural slant to notice novel information, allowed her to attend to subtleties imperceptible to others. She could then assess the information in the context of her existing memories, which allowed her to recognize a pattern as the brain dictates. She attributed her own meaning to the pattern, which she translated into action with a positive outcome. Whether the experience was mystical or nothing more than a lucky presumption becomes less relevant. Regardless of how her actions were derived, she saved a life. And any act of healing is miraculous.

These experiences are remarkably similar to the sensory illusions reported by nonsynesthetes when using LSD. In a functional MRI study published in the *Proceedings of the National Academy of Sciences of the United States of America,* Dr. Robin Cahart-Harris at the Imperial College London found that use of LSD was associated with an increase in brain activation of frontal and parietal cortex in a pattern that interestingly resembles activation patterns in infants. Frontal and parietal cortex functions are tied to social cognition and processing sensory information. If the networks in these regions activate in harmony to help us construct a sense of "self" and a sense of our external environment, then it is possible that activation in even greater synchrony might also allow the capacity to somehow blur these perceived boundaries, just as mirror-touch synesthesia does.

The unconstrained flow of information between the networks of these cortical areas is also associated with a blanketing sense of peace and calm, what some call "subjective well-being." This may be the case in synesthetes who perceive the experience of dissolution or oneness as egosyntonic. But, if synesthetic experiences are egodystonic, synesthetes may instead try to reduce the distress from their sensations through coping techniques, like avoidance. Habitual avoidance is unsustainable over the long term, however, because it leads to extended periods of isolation and a persistent, heightened, and unregulated state of hypersensitivity.

One mirror-touch synesthete, known as Amanda, avoided her synesthetic experiences so diligently that, without realizing it, she eventually withdrew almost completely from the world outside her home. Amanda became a recluse. For a larger part of her life, she minimized contact with others outside of her home and even avoided eating in front of others, including her family, to avoid the mirror-touch experience that occurs for her from seeing others chew. She could not bear the thought of bringing a dinner table into her house, let alone sit at one with others.

Fiona pushed her chair back slightly and glanced under her desk. "What's that? Are you hungry?" She bent down, and like a Las Vegas magician, pulled out a large fluffy black rabbit. She cradled it in her arms and hand-fed it some kale while stroking its side. Despite Fiona describing herself as mostly isolated, I considered that perhaps she had found the social connection she needed with her animals, as if they had been an integral part of caring for her as much as she cared for them.

"My animals, they connect me with the universe," she said. "I love to watch them. You know that feeling you get when you're in a lift and someone presses a button and the lift starts to move up or down? There's that shift, especially when you're stopped. Well, when I watch birds take off in flight, I get that feeling. That lift. It's a shift in your gut. When I feel isolated, it connects me with a sort of higher thinking about the universe where isolation doesn't exist."

Fiona began to stroke the head of the rabbit. Holding her rabbit reminded her of how her mother used to stroke her head. She would fall asleep in seconds. Stroking the bunny, she was also stroking her own hair, her own head. She confided that as much stress as there can be from her synesthesia in general, she can also rely on her synesthesia to bring her to a place of calm. "Watching leaves blow in the wind, I might have the same feeling like when I watch birds fly. Or the touch of brushed cotton or the touch of my dogs' coats—there's a certain sweetness to them."

Fiona was confident that her synesthesia was at the heart of her connection with animals. She began grooming her pets in her garage. Neighbors noticed the delight in her care with the animals. Soon all the people in her neighborhood brought their animals to her. Fiona had a talent with animals. She knew there were different ways to hold animals, certainly avoiding overextension of limbs, but she also noticed that there were some positions for each animal that worked particularly well. "I've gotten quite good at figuring it out," she said with a grin. "It's weird. I haven't done any vetting science. I haven't done any specific course, but a lot of it is just intuitive and it's worked well so far." She's developed a talent for removing fleas as well. When she sees the fleas on the dog's face, she gets a "creepy crawly" feeling on her face and uses her own mirror-touch sensations to guide her hand. She also takes immense pleasure with animals that are ticklish, especially Yorkies, grooming and giggling together in unison.

Grooming a big, happy long-haired German shepherd might require intense work, but Fiona feels a sense of bliss afterward. This is the true reward for her labor. Of course, there were some dogs who were tense, which drained Fiona. When they tensed their bodies, she felt as if her body also tensed, both standing there almost taut. But, in these tougher times, she is driven to find what calms the animal, to locate that itch that needs to be scratched. She couldn't groom one dog's face until she figured out that if she put the dog in water, she became putty. Another one, as long as he was on his back on the floor, she could do his nails, brush his fur, anything. Being attentive to the anxiety level of her dogs was key to her talent. She reflected, "The dog picks up on my anxiety levels as well. I'm very aware of that." She found it similar to her interaction with human beings where she could sense the feelings of agitation or happiness or sadness in other people. Sometimes it would be just as simple as a matter of playing soothing music. She found that she could tune in with both animals and people, read them, feel them, then meet them in the middle. "By calming myself, I calm them, too. And vice versa. It's part of the process."

Fiona's sense of connection extends through all species, but she noted that birds have the ability to fill her with the greatest wonder. "Their twittering and chirping, it's like they're dancing through levels of sound and I feel the dancing in me. It's just unreal."

I asked Fiona if she had found a way to harness her experiences, to perhaps handle stress when it happens in her life. She told me about an incident a few nights earlier. It was a night like any other, but the street was poorly lit. She looked in the rearview mirror and saw an empty road. She pulled up in front of her home and switched her lights off. She then opened her car door and a bicyclist rode right into it. And at the exact moment that he crashed into the door, Fiona had an immense response. She felt as if she had been rammed against a wall and hit in the back of the head, attacked by an invisible force. It was only after she was able to look down that she realized that there was a man on the ground. At the same time she felt as if she were him, lying on the ground. He stood up in a rage. She could obviously understand why. They were both overwhelmed with distress. He sputtered a few more curse words before picking up his bike and riding off. Fiona came into her house, still shaken, still sharing in his pain. He rode off. There was no closure. However, Fiona's self-awareness is superb. She knew she had to alter her sensory experience. She walked over to her bunny and lay down, stroking its head. She felt as though she was being stroked as well, even if just through her mind's eye—catharsis through kindness. "It's very hard to explain but it's still just as real for me." She looked up as if she had seen a thought fly overhead. "It's just one of my ways of managing my mental world."

Fiona told of another time when she was walking her dog and ran into her landlord. He casually mentioned that because he was moving out of the country she needed to either buy the house she was staying at within six months or move. To Fiona, this might as well have been a bomb. Her greatest fear was homelessness; even the possibility was terrifying. She felt a primal anxiety rise up in her. She felt she couldn't breathe and the sensation that her body map was distorting, her legs rising while her arms melted into her gut. Rather than losing herself to the feeling, she walked away and took a different route than usual. She found an area with many trees. Freshly fallen leaves carpeted the ground. She let her dog off the leash. She focused on only watching her dog's movements. By watching her dog's body move, she felt her own body move. Her limbs experienced the physical sensation of trotting along among the leaves. She felt her body map steady and gradually radiate outward. She moved with her dog by being still, observing quietly. She was once again able to think calmly and realistically, able to sort through the worst-case scenario and plan her next steps logically.

Fiona paused. I felt the sensation of her hands folded against her upper abdomen. "To manage these situations when they come up, I really find it most helpful just surrendering myself *to myself*." She pointed her thumbs upward at me before saying, "You have to surrender to yourself to accept. I see it as acceptance. I see 'surrender' as a positive word. There are some things that you can't change or you don't have control over, and there are certain things that you don't *need* to have control over. You just kind of let it be."

CHAPTER 7

Holding Up a Sign

IN THE WEEKS FOLLOWING MY CONVERSATION WITH FIONA, I continued to reflect on what it would mean to surrender to the self, especially when it seemed that my "self" was created by other people, the narrative that synesthesia is a *condition*. If I did consider synesthesia a condition or a disorder, would my only "treatment" options be dulling my senses or cutting myself off from others? Though in doing so, I would also prune away at the human experience, by twig and by sprig, until all that is left is a clean, lifeless stump. I knew I was ready to surrender, but not to this cause. Instead, I was ready to surrender to the narrative of a greater purpose. Greater than pain, greater than fear, and far greater than the temptation of mindless comfort. Liberation meant surrender. I had to surrender myself to *my* self. And yet, I had entered into medicine with the belief that surrendering myself to others, to the field, was my purpose. I had entered my marriage with the belief that surrendering myself to my husband, to the institution of marriage, was my purpose. But I was mistaken. In truth I had been attempting to live my life through the reflection of an aimless martyr, all too willing to let myself be erased.

Throughout my medical training, I had continually searched for that elusive—and ever-daunting—sense of purpose in the places I thought I knew best, in structures and institutions, in my spouse, anything other than me. Connecting with Fiona and others, however, helped me realize the *what* and *why* of purpose is not nearly as vital as first understanding its *where* and *who*. While I continued to drift in my personal life, growing more and more alienated from Jordan and quietly losing touch with my family, especially Rainier, I was stunned to discover that, at least professionally, I was finally brushing up against the warmth and coolness of purpose, almost by accident. I accomplished this not because I blindly offered myself in sacrifice but because I had started to cultivate my own sense of family to include an array of others—other residents, other synesthetes, other stewards of atypical brains, other tribes.

We are all, in our own way, in search of our own tribes, a group of people who understands us and helps us find peace. Our most authentic sense of purpose is likely to be found by growing our family, our community. "Family" extends beyond people with whom we share a blood or legal bond. Family is also our network of support, including friends, colleagues, collaborators, fellow thinkers and feelers. Family is made up of individuals who care for each other and help one another become a better version of their truest, most liberated self. Ludwig Wittgenstein proposed that family resemblance transcends the biological and can consist of the fluid similarities and affinities that flow between us. Our connections form most easily with people with whom we share the most values and experiences, thoughts and feelings. Thus, we are the family that we become just as we become the family that we are.

However, humans also have the capacity to expand our perception of what we share, what connects us. While I can appreciate that each synesthete has a diverse set of synesthetic associations with varying intensities, I can also appreciate that to be "understood," I actually do not *need* to find other synesthetes with the exact same synesthetic associations as me. I can find even more meaningful affinities outside of the innumerable minutia of my sensory experiences to connect with synesthetes and nonsynesthetes alike. I only need to be understood *enough*—just as I only need to understand and immerse in the feelings of others *enough* to connect, to share in their experiences without losing myself.

Fiona and CC taught me that through self-awareness and a nuanced awareness of human presentation, I can appreciate that we are unlikely to ever find others who experience an exact replica of our own sensate world but that we can find an almost endless number of people who share an affinity in at least one of our dimensions, members of our most genuine family. I realized that, rather than avoiding others because of my synesthesia, my synesthesia has given me a rare opportunity—a gift—to find deeper meaning, as a human and a physician, in the liminal space between the "little self" and the "big self" as well as allowing me to perform the great leap between thinking and feeling to access a deeper understanding of empathy, family, and community, the tribes that define us and, for better or for worse, shape us.

Where then does such a family begin? Are we and our stories automatically born into a specific tribe, or is our tribe born through us, through the stories that draw people closer? While I was able to get a better handle on my trait professionally, I still struggled with it personally. Seeking out the answer to these questions, I hoped to learn what were those critical pieces that I was missing—what I had missed in my relationships with Cristina and

Jordan, what I had missed in creating my own tribe. I needed to know what had prevented me from making the type of connections I most craved, the kind of emotional connections that help myself and others thrive. It was around this time, as I silently wondered about my family and community, about human connection, that I came upon Rosie and listened to her story.

At 5:30 A.M. on March 25, 1997, Terry Doherty gave birth to Katherine Rose Doherty, her fourth child. She weighed 7 pounds, 3 ounces, measured 21 inches long, with an occipital frontal circumference of 37 centimeters, measured across the back of her head to the tiny ridge above her eyebrows, which were slick with sweat. Her first picture shared the harsh exposure of a hospital mug shot, though softened somewhat by the ruddy cheeks of a newborn baby girl with glinting dark brown eyes, wrapped up delicately like a petite rosebud at the end of a winter thaw. She emanated warmth. Almost immediately, Terry and her husband, Mike, started calling their newborn daughter Rosie.

Terry beamed for weeks the way only a seasoned mother can. She knew the routine. She excitedly tracked Rosie's growth against the typical developmental milestones. Ten fingers, ten toes at birth. She latched easily to Terry's breast. At two months, she cooed, smiled, responded to sounds. At four months, Rosie babbled, rolled, held up her head. Two months later, Rosie sat, stood, rocked and, three months later, at nine months, she started to walk, months earlier than her brother and two sisters had started pattering around the carpeted floors of their humble southern Massachusetts home.

But a year into her young life, something strange happened. Rosie seemed to babble less and, with each passing month, paid less attention to her adoring, attentive sisters. She seemed only interested in walking around the house wildly banging things together—blocks, utensils, books, anything she could get her tiny hands on, ignoring or oblivious to her family. Terry and Mike waited for Rosie to say "mama" or "dada," any simple declaration of love or need to bubble forth from her cherubic mouth. But their anticipation quickly turned into a quiet anxiety, a terrible desperation. Rosie drifted out of Terry's maternal reach and into the unknown. She was suddenly alien. Terry felt as if a switch had been flipped off in her baby.

When she turned two, Rosie skipped running. Instead, she bolted feverishly around the house. She was frenetic and tireless. She tore her clothes off incessantly. No ruffled pink dresses, no shirts, no shoes—not even pajamas. She couldn't tolerate clothes of any kind. While other children were slowly getting the hang of hopping on one foot, Rosie started finding ways to leap out of windows. In one instance Terry had to chase a naked, howling Rosie around the cement driveway for over twenty minutes. Was this her child, she wondered, or had her daughter been spirited away, leaving in her place a wholly unknown species of animal, a savage invader in their home?

"I knew something was wrong," Terry told me, "but to convince the pediatrician there was something wrong was just as bad." A nurse by training, Terry always arrived at the pediatrician's office on time, hoping to get an explanation for Rosie's troubling behavior. But after waiting in a room full of coughing toddlers and children, she would always leave a few hours later with reassurances that Rosie would be just fine. *"Oh, she's in a big family, that's probably why she's not talking much."* Rather than taking comfort in such reassurances, Terry found them unsettling. She knew there had to be someone out there who would listen more carefully to her concerns about Rosie. The pediatrician wasn't being careless; the diagnostic red flags that would now almost undoubtedly catch her attention were not readily available at the time. With no other options, Terry turned to the Internet, which in 1999 held only sparse archipelagos of information. But her online searches turned up the same word again and again: autism.

After a year of Terry's persistent pushing, Rosie, now four years old and unable to mutter a single syllable, finally arrived at an arena assessment, an evaluation of sensory and motor functions. Without the benefit of blood tests or brain scans that can provide a definitive diagnosis, arena assessments were designed to emphasize observation of behavioral nuances by multiple clinicians. It had taken a year to get the appointment at their local children's hospital. Terry came in on her fortieth birthday. Hours went by as Terry struggled to keep Rosie under control while a team of specialists—a speech pathologist, an occupational therapist, a medical-education specialist, a social worker, a child psychiatrist, and a developmental pediatrician—took turns observing Rosie. Too many names and faces for Terry to recall, but the last ten minutes of the assessment remain far too clear in her memory. Terry will never forget this moment. Sadly, neither will Rosie.

Under the pressure of their collective gaze, they brought down their judgment, "Your child has autism." Then there was silence. Terry waited, patiently. While this did finally provide a concrete diagnosis for her and Rosie, she had already come to this conclusion on her own and was surprisingly no longer bothered by this now-familiar word. But what came after continues to sting. "She is going to be your child for life and she is nonverbal, profoundly developmentally delayed, and profoundly mentally retarded. She is never going to achieve what you may have hoped for."

"I'm not really a crier," Terry confessed, "but I was crying." She labored her way back through the children's hospital, past endless murals of simpering celestial bodies, tigers with Cheshire grins. Standing in the crowded parking garage, she struggled to get the keys into her car door. She sat in the driver's seat with Rosie on her lap, the jagged set of keys clutched tightly in her right hand, digging deep against her palm. She dropped her head forward over the stitched leather steering wheel and came undone. Certainly, she thought, these specialists had seen so many cases that by now they could see her future so clearly. Perhaps it was their place, their responsibility, to bring right up against her face the vision of a future that she thought would resemble hope but instead only promised despair.

Terry drove home defeated, repeating over and over again the words. *Profoundly developmentally delayed. Profoundly mentally retarded. Profoundly nonverbal. Never.* As she sat at the kitchen table, clutching her rosary, she tried not to settle into the cursed narrative of a cursed mother with a cursed child. She would have easily slipped into that darkness forever, she told me, if she hadn't noticed behind her the sound of a wet metallic clunk. Followed, after a pause, by another heavy clunk. And then another. Terry slowly raised her head. Rosie was standing atop oversized juice cans that Terry had left on the kitchen floor. One after another, Rosie had stacked together six cans to form a small set of cylindrical stairs, which terminated a few feet below the lip of the kitchen counter. To Terry's surprise, Rosie was hoisting her right leg onto the counter, precariously balancing on a wobbly can, her small left foot gracefully on point. Before Terry could reach out to grab Rosie, a voice entered her mind. The voice reverberated with all the maternal force of Shakti, the primordial Great Divine Mother, and told her to wait, to *surrender* if even for a second, and let Rosie climb. She remembered thinking at that moment, "Exactly *who* am I trying to fool here? This is the girl that, even though she is nonverbal, climbs up cabinets. This is the girl that, even though she was supposed to be profoundly mentally retarded, is thinking. She is problem solving. This is the girl who saw me open the windows over a hundred times and figured out every step she needed to take to get out the window. She pulls the windows open herself. You can't tell me she's *that* profoundly developmentally delayed, right? She has obstacles, absolutely, but this does not have to reflect who she is as a person. Everybody can have a bad test day, right? Just because doctors fail their boards one day, does that mean they're not going to be great doctors the rest of their life? Absolutely not. Today we failed a test. That does *not* mean we will fail every day thereafter."

Terry wiped her face with her sleeve and walked over to Rosie. She lifted up her daughter—bare feet, tussled hair, and all—and simply held her closely against her chest. "We're going to work on this, Rosie. We're going to do this."

Terry's first step was to educate herself. She waded through as much knowledge as she could through books, websites, online communities, and support groups. Whether the information was rigorously backed by science was not as relevant to her as the potential insights she could glean from new perspectives and ideas. Still, she was skeptical. Partly through her instincts as a mother and partly through her experience as a nurse, Terry understood that what works for one child doesn't necessarily mean it will work for another child. Only Rosie, Terry ultimately understood, would be able to determine whether a potential therapy was successful.

At the core of Terry's strategy was the idea of a "super team," one that could help her determine a viable way to help Rosie. She recruited her two other daughters, Elizabeth and Christine. "They were like my stooges." She gave them assignments to complete with Rosie and suggestions about how to help Rosie focus and work with them. Elizabeth and Christine had free rein with Rosie in this regard, provided that whatever they came up with together helped Rosie. Meanwhile, Terry employed more conventional tactics with Rosie. She put up a large flashcard featuring pictures of objects or actions with their corresponding names tacked above them. She spent hours upon hours flipping through pictures for Rosie, saying the name of the picture out loud, pausing in silence for a few seconds, repeating the word, pausing again for a response that Rosie refused to, or was unable to, volunteer, then moving on to the next picture.

Bank . . .

Baann-kk . . .

Coffee . . .

Cawh-ffeee . . .

Some days Terry felt as though she were simply repeating these words out into the atmosphere with no sign of retention from Rosie, but she persisted. She printed out more pictures, laminated them, and placed them around the house with the hope that Rosie would stumble across them and recognize the sound of the word with its image and written representation. When she was potty-training Rosie, she taped up pictures of toilets, bathrooms, and toilet paper in every room of the house. Terry paid close attention to the objects around the house Rosie gravitated toward most, what most often caught her attention, and where she spent most of her time.

This was most often the imaginary town of Princessville, the capital of Rosie's upbringing, populated by Disney princesses, Rosie, and her two sisters. Every day, Rosie and her sisters created a new drama. There would be debacles and kidnappings that would last for weeks with extravagant backstories that became progressively more elaborate. A common theme was their re-creation of an episode of *The Bachelor.* Every week, twenty-four Barbie dolls competed to win the heart of Prince Charming (Christine's Ken doll). Rosie participated in all of the juicy drama. She played with the princesses or prince, and together the sisters staged birthday parties, galas, and balls. They created the *Princessville Bulletin* on a whiteboard, which was covered in pink dry-erase markers with hearts of every shape, size, and configuration. They listed announcements with dates and times of upcoming events and a list of invitees. Timely blow-out sales at the Princessville beauty salon occurred regularly. They erected princess schools, hospitals, and town halls. They declared and celebrated Princessville government-sponsored holidays. For every scene, for every holiday, they made ornate backdrops and intricately involved decorations, which were handcrafted out of white paper cutouts and Crayola markers.

From time to time, Terry stepped in to direct Elizabeth and Christine. "If we needed to work on engagement and eye contact, I would tell them, 'Okay, so this is what you're going to do, and let's just see what happens.'" When I asked Elizabeth if they felt that they were forced or were making a deliberate effort to help Rosie, she replied, "I think it was mostly just having fun, but maybe as we got older a little bit we recognized it was helping Rosie. Even then, it was just fun."

Unlike Terry's educational efforts and sessions of forced repetition, Rosie's visits to Princessville seemed to make the greatest difference. The magic Rosie found there is no mystery to Dr. Adele Diamond, a leader in childhood brain maturation research at the University of British Columbia and a pioneer in the field of developmental cognitive neuroscience. Her research has steadily evolved to hone in on the importance of activities like dramatic play in childhood brain development and its influence throughout the rest of a child's life on healthy neuropsychological functioning. Many of the same concepts form the basis of Montessori education. The absence of structured or external rewards, for instance, provides children with internally motivated rewards: learning and mastery. Children learn to help each other, take care of each other's environment, and take turns teaching one another, providing hidden forms of tutoring that avoid embarrassment or punishment from an authority figure. Rather than excluding Rosie or playing with her as if she were incapable of participating, her sisters played with her just as they would with any other friend. She began to recognize patterns in the stories that Christine and Elizabeth laid out. As a result of their play, Rosie was able to more closely interact with others and for longer periods of time. Terry noticed it became easier to introduce Rosie to new people, to bring her into new environments, to gently push Rosie close enough to the edge of her comfort zone so she could step out of her internal world on her own.

A key component behind the efficacy of dramatic play is in how it can support development of executive functioning while providing early education in social concepts and skills, which makes the highly contextualized nature of dramatic play all the more relevant. Elizabeth and Christine took advantage of this and were able to weave into their play scenarios that were relevant to Rosie's daily life. When Rosie's class had a scheduled field trip to a museum where the highlight would be an authentic Japanese tea ceremony, Elizabeth and Christine came together and played a game with Rosie called "Cinderella and the Tea Ceremony."

A few months later, because of "Snow White and the Massachusetts Comprehensive Assessment System," Rosie was able to sit through her standardized academic testing at school. An unbelievable feat considering how she could barely sit on the couch for five minutes let alone a tiny classroom chair for hours.

Being mindful of Rosie's immediate context helped make the stories engaging, applicable, easier for Rosie to incorporate into her long-term memory. Rosie loved Spider-Man, so Spider-Man (as played by Rosie Doherty) became a lead character in many of the stories—from "Sleeping Beauty" to "Green Goblin's Day in Court." Rather than forcing lessons, Terry understood the importance of genuinely "taking their interests and weaving in what you want to help them learn."

These social stories provide a structured method for describing a situation, skill, or concept. The basic history of *social stories* extends far back and has ties to several intellectual figures, including Lev Vygotsky, the developmental psychologist who proposed in the early twentieth century that development of higher cognitive functions in children emerges through simple activities in a social context. Vygotsky's scientific legacy influenced the work of his student, Alexander Luria, one of the fathers of modern neuropsychology. Luria's writing would later play a central role in Oliver Sacks's decision to study the brain through the lens of a naturalist and the tools of narrative detail. Social stories have anecdotally helped many in the autism community cope with the anxiety of the unknown and navigate socially complex scenarios. However, there is more at work here with tangible implications for all of us. I had been so single-mindedly focused in my clinical training on memorizing the minutia of medications and sterile procedures that I had ignorantly dismissed the most ancient of human tools: beyond puppets, beyond dolls, beyond words, the raw power of *storytelling.* Despite all the technical knowledge I had accumulated in medicine and the sciences—stacks and stacks of scholarly tomes filled with pithy facts piled high—it was not until meeting with Rosie and her family that a humble fragment of insight connected me back to something unmistakably primal yet uncanny in its familiarity.

Sharing stories can help us to heal, grow, and thrive. They have the capacity to touch our internal psychological state and thus our physical bodies, too. In this sense, the psychological is also biological. Whether we are aware of it or not, our personal narratives are constantly shaped by ourselves and others.

To make Rosie's social stories richer and more effective in reducing anxiety, Terry investigated locations that Rosie would be visiting for the first time. Extracurricular espionage was in addition to Terry's full schedule as a nurse and a mother of four. Yet somehow, she managed to scope out new parks, new summer camps, new petting zoos, new classrooms, new doctors' offices. "I seriously felt like Rosie's Secret Service."

Terry was driven by the hope that Rosie was, indeed, understanding and learning through delicately crafted homemade stories. Mindfully assuming that children, including autistic children, are capable of authentic emotional relationships is a premise also present in several other behavioral interventions for autism. Though, the scientific evidence behind behavioral interventions remains controversial. Invested developers and pleased proponents defend their choice of interventions while concerned skeptics and disappointed parents find some interventions overblown in their claims and even financially exploitative. However, it is important to acknowledge that the evidence for many interventions is neither conclusively for nor substantively against the hypothesis that they have a meaningful clinical effect in children with autism. The field is wrought with mostly unavoidable limitations in the development of interventions and how we study them. There are sufficient challenges in simply attempting to define a "meaningful" outcome.

For Terry, this meant anything, *anything,* that moved Rosie forward, even if only a little. The intervention may not be generalizable. For example, the intervention may show some effect in experimental settings but also may have been unintentionally designed for a specific subtype of autism present in only a handful of people. Or perhaps the intervention has ambiguous results because the intervention was developed through small-scale pilots in a particular setting, such as nuclear families from middle-class neighborhoods in an ethnically homogenous locale of the Midwestern United Sates. Therefore, caregivers and researchers may benefit most by maintaining a rational balance of openness and critical appraisal of the available evidence for behavioral interventions.

Helping an autistic child learn to navigate his or her internal and social environment is likely not going to be accomplished by one intervention but rather a series of orchestrated attempts working in parallel to affect several contributing factors. Rosie may not have been maturing at the same rate and sequence as her peers, but she was still maturing, learning, and growing. Because the methods for assessing development in a child on the spectrum are limited, a child will typically continue to fail conventional tests and is almost always underestimated. Thus, we may never know the exact details of Rosie's neuropsychological development. But what we do know is that Rosie's family was persistent, compassionate, open-minded, and whole-heartedly mindful. Likely because of these integral human factors, Rosie can now share insights from her experiences growing up—in her own words.

"I wanted to start to talk to people my whole life," Rosie said. She was dressed in an oversized pink puffer coat. As we spoke, she sharply turned her head to stare at an empty corner of the spacious conference room then whipped her head back to face me. She emanated a spherical field of crimson 2s with magenta highlights, assertive and feminine, lined underneath with shy violet 3s. I felt a light brushing on my eyelids as her short chestnut hair swooped into her face every time she turned her head. She looked toward me but not at me—more than just a semantic distinction, though perhaps less relevant than her act of deliberately attempting to display genuine interest in another human being. "Not being able to talk was the most frustrating part of my life!" Her torso shifted toward my general direction. I could feel a genuine rise in emotion infused throughout her words and in her body. This was a young girl who was fully grateful to finally have an opportunity to speak and was now pouring forth language with every breath and every muscle. She radiated joy, waving her arms in short punctuated gyrations that had the mirrored sensation of tiny fireworks bursting on all of my joints, like sparks of gratitude. "I was trying to tell people how I felt, but I couldn't. I couldn't tell *anyone* how I felt!"

For much of her childhood, Rosie recalls strangers regarding her as incapable, frozen in a state of eternal disability. Somehow the perception that she was incapable of speech was conflated into the assumption that she was likewise incapable of thoughts and incapable of deep emotion. Rosie has a clear memory of when her mother was told that she was "profoundly developmentally delayed and profoundly mentally retarded." Rosie refuses to include the words "mentally retarded" in her now rich vocabulary. Through all of her tantrums with others and her impatience with word drills, she explains one root of her frustration and her resistance to these efforts. "I just wanted to tell people how I saw the world. But I couldn't."

And one day, like a monsoon gingerly announcing its arrival, she released a single drop of language: "Rosie." Thereafter came two: "Rosie want." Then three: "Rosie want juicy." Then enough words to describe detail: "orange juicy for Rosie." And then, a torrent of language unceremoniously rushed forth into the world. Terry marveled at the ongoing deluge with a level of excitement comparable to Rosie's. "This girl who I was told would be nonverbal for life became a chatterbox. We actually have to remind her not to keep interrupting people. It's incredible. The way she learns is definitely different than other people. With most kids, you see them progress step to step. With Rosie, you don't see any progress and then there's this leap. You can practice for a while and nothing. Then all of a sudden you see all the progress at once. It took me three years to teach her to learn to ride a bike. Think of all the motor coordination in that. Riding, steering, peddling.

I had numbers on her socks: 1 on the left, 2 on the right, telling her 'one, two, one, two.' You never think about how much there is that goes into things until you try to teach somebody."

Terry would help push Rosie's legs while she held desperately on to the handlebars of the bicycle with her tiny hands. "Nothing, nothing, nothing, and then all of a sudden she took off one day. Boom. Done. Same way with reading."

Rosie chimed in, "My second-grade teacher told me and my mom, 'Give up, she'll never learn how to read.'" Without skipping a beat, Terry brought her hands up to her temples in disbelief. "All of a sudden, we're reading Shakespeare."

Recently, Rosie became engulfed by the events surrounding the life of Alexander Hamilton. "I love reading about history. Alexander Hamilton didn't have a boring life." She began to pick up momentum and her gaze began to loosen back towards the corner of the room, "Alexander Hamilton had a very interesting life. So, you're like, five pages in, and you're absorbed in that book. What is he going to do next? Now he's the governor. What is he going to do next? You always want to know. Anyway, and then you're like, he spoke Greek. Really! And what is he going to do next *now*?"

I suspected that had I allowed Rosie to continue, she would have been able to deliver an entire graduate course on the life and times of Alexander Hamilton interrupted by only a few deep inhalations and perhaps one bathroom break. Curious, I asked her about what her experience was with sensations when she was younger, mostly to see if her tactile hypersensitivity improved as her language improved. "My senses are *mezza mezza* I would say. Because, when I was younger, I liked to stand in the snow with my bare feet all cold. I also once sat in a bathtub that was wicked hot, and I couldn't tell it was really hot. Christine had to take me out of the bathtub. I was really sensory. I still kind of am, but not as bad as when I was younger. Clothes irritated my skin, so I hated wearing them."

Elizabeth, the eldest sister and now a practicing critical care nurse like her mother, gleamed with a sheepish grin. Her face was framed by red hair and freckles accompanied by a bright red 2 in a nest of wiry tangerine 5s laden with dollops of yellow 8s. "She didn't want to wear horseback riding pants one day. She eventually put them on, but she refused to change out of her nightgown and we were running late. I finally got sick of arguing so I said, 'Fine, wear the nightgown.'"

"Well, I hated the seam that ran from the front to the back. It felt terrible. Elizabeth ended up having to stuff the nightgown into the horseback riding pants. I looked like I was a pregnant fourth grader riding horseback."

Terry bemoaned, "It was a big fight. Every Sunday."

Still riding her own momentum, Rosie trotted elsewhere in her mind. "To help when I was younger, mom would sometimes take a brush and she would brush my skin."

"You know, the brush with the soft bristles to desensitize," Terry explained. "It's called Wilbarger brushing. It's supposed to actually heighten the feeling on your skin so you know where your body is in space. In autism, they just don't seem to make connections well between the nerves in the skin and the actual sensory part of the brain. So, you make them aware with the brushing. You can also do joint compressions and cranial pushing so that they can realize, 'Oh, here I am. Here's my body.' And they can tolerate a lot more. We even had a weighted vest that we used a few times and a weighted blanket. I had forgotten about the weighted vests. But I tried to make them for whatever her preference was so she would actually wear them. I think I must have made 500 weighted vests. We were always changing the bean bags inside of the vests around."

Rosie dug her hands deep into her pockets and leaned forward over the table. I could sense the delight in her nostalgia as a memory materialized. "I had a butterfly vest. It was my favorite."

Terry smiled with contentment. Terry had a heavy maternal blend of merlot 2 balanced by a flush of deep, almost cobalt, 4s. She paused in this warmth for a moment and then calmly continued, "Eventually the sensory defensiveness became less."

Challenges with sensation are common in persons with autism spectrum disorders. They may be hypersensitive to a tag on the back of a shirt while hyposensitive to a cut or scrape. I can relate to the insufferable scratching from almost any tag in my shirt, particularly as a child. Although I can't say that my hypersensitivity has abated, I've become much more deliberate about adapting my environment. For as much grief as it caused me as a child, all it takes is the decisiveness for a single snip to give life a little more ease. It's a simple solution with virtually no harm to myself or others to cut off tags that even hint at bothering me, if I haven't already avoided them altogether—a minor act of kindness to myself.

In autism, extreme fluctuations in sensitivity exist in virtually all senses: sight, touch, smell, taste, hearing. It even affects an autistic person's sense of equilibrium and positioning of their joints. Why this is the case is not yet known. Some studies suggest that sensory overload or conversely a dampening of senses may be related to three factors: hypersensitive peripheral nerves; disordered high or low activation in corresponding sensory regions in the brain, potentially linked to gene mutations; heightened activation or connectivity with neuronal networks involving the insula, the part of the brain associated with experiences of disgust or aversion. One practical way to address this challenge is sensory integration therapy for autistic people, which involves the use of self-reflective play activities to bring awareness to sensory stimulation to help gradually change how their brains react to the activities. While such methods had been controversial, a 2013 study finally provided some evidence in support of what many parents had anecdotally reported.

"Another thing mom would do was the brain gym. I still do it, too." I assumed that this was some sort of mainstream brain training game that had inevitably been transformed into a novelty smartphone application.

"To calm herself down when she gets really nervous, I might ask her, 'Where does the tongue go? Where is it?'"

"At the top of my mouth," replied Rosie, reliving the memory of her mother's attempts to soothe her.

"We would also work on movements that went across the middle of her body because she had a hard time with it. To help her focus her attention on the sensations of her body in space, she would cross her arms over, hold her hands, and then flip them over and try to figure out which hand was the right and which was the left."

"I started doing brain gym at school in about fourth grade and it was pretty good. But I didn't tell anyone that I was doing it until about ninth grade. When I got upset with something or someone would bother me, I would go out in the hallway and do it. Some people didn't like me going out in the hallway, though. They would tell me, 'You can't do that!' and I would tell them, 'Yes, I can. I'm trying to calm myself down. I don't want to be anxious.'"

Terry leaned back and I had a glimpse of the kind of compassion she had so carefully cultivated in their home. "It's hard to have a different brain in a mainstream world sometimes. Isn't it, Rosie?"

"Yeah, it's really hard. Still, people will dislike what you do sometimes, and you just have to sometimes accept what their opinion is and move forward and then continue doing whatever you need to do, but tell them in advance the next time. It's hard being in my world where you have autism and people think you're not a normal person and you're not treated like a normal person. You're treated differently than everyone else and it's hard. It's really hard. But I keep moving forward. I keep being positive, and I don't let people judge me. If they judge me, if that's what they think, then that's what they think."

"Where would you feel most normal?" I asked, wondering where this anchor came from, where she had the opportunity to learn how important it can be to expand one's own definition of *normal.*

"Around my sisters. They were so *awesome.*"

I noticed that the word *awesome* had unique qualia. It was a genuine, crisp, spritzy, cornflower blue that overpowered the typical synesthetic colors of the word *awesome,* which I usually experience as a more muddled combination of green, red, orange, white, and brown hues. Rosie infused the word with a deeper sense of awe and appreciation than what I was used to, synesthetically speaking.

The entire family was a team. They deliberately integrated the ways to help Rosie into their lives as a way of avoiding the kind of burnout often seen among caregivers.

Social support is one of the key functions of a successful support network. There are generally five dimensions of social support: instrumental, informational, emotional, affirmational, and belonging. Also known as tangible support, instrumental support consists of how others address our most practical needs such as assistance with chores, grocery shopping, or opening a jar. Informational support is how we learn or find information about the world, which can manifest in the form of receiving advice from a friend who is your own personal fashion or technology guru. Emotional support involves having someone available to provide you with the feeling that you are understood and accepted regardless of your circumstances. Affirmational support includes the love and affection from others that helps us feel valued or respected and can be a source for cultivating self-esteem. Belonging, or companionship, is the type of support that helps us feel connected to something greater—a community, an organization, or a family—and can have some overlap with emotional support. Belonging may manifest as simply having someone available to listen to you when you need to talk. Dr. Lisa Berkman from the Harvard T. H. Chan School of Public Health studied the impact of social supports extensively in her Human Population Laboratory study. Over nine years she followed 6,928 adults in Alameda County, California, and found that people who lacked social support and ties to their community were more likely to die regardless of age, socioeconomic status, and self-reported physical health.

In my own work I examined the dedicated participants of the Framingham Heart Study, which has closely monitored more than 5,000 men and women and their children since 1948. This humble, closely knit community has helped provide a sizable portion of what modern medicine knows about cardiovascular health and other health conditions. I was interested in investigating what we could learn about the association between social relationships and the risk for stroke and dementia. It turned out that those who had the greatest availability of someone to provide emotional support and a sense of belonging to a specific family or tribe had a significantly lower risk of developing these devastating chronic brain diseases. Our findings confirmed some of what has been seen in terms of the importance of social engagement and supporting brain health as well as the harmful effects of social isolation. One key finding revealed that people with the greatest availability of emotional support were more likely to have higher levels of brain-derived neurotrophic factor (BDNF), a brain molecule critical in repairing neurons and creating synaptic connections between neurons. The creation of neuronal connections is central to neuroplasticity—the brain's capacity to change its many networks and synapses in response to conditions inside and outside of the body. These results hold promise in helping us understand how the social and behavioral aspects of our lives can influence the biology of brains and, ultimately, our minds, our behaviors, how we connect with, support, and influence the lives of *others*.

With Rosie able to communicate with greater ease, she was able to share and explain how she thought and processed the world around her, which helped her family tailor their interactions into a language (alien to what most consider language, but now a native dialect of the Dohertys) that Rosie could understand.

"They say I'm a visual thinker, so thinking in words and numbers doesn't really make sense for me. If I think in words and numbers, it makes everything more confusing."

Recalling how unaware I was that my perceptions and interpretations of sensory information were at all different from others, I wasn't sure what to expect when I asked, "Rosie, how do you think about people?"

She took in a deep breath as if getting ready to dive, "So, my brain has consisted of these balls and each person has a color, a colored ball. Okay? And the black one is the only one you don't want. The black and the white. You don't want to be a black ball." She moved her hands in front of her, visibly manipulating her mental representations. "You're a green ball, okay?"

I was tickled by the coincidence that my favorite color is green, forest green with hints of sea green to be specific, like the green waters of a Pacific shoreline.

But what I found most surprising about Rosie's account is that Terry, Christine, and Elizabeth were unflinching. They had been so mindful of Rosie's perspective by now that they were able to embrace this aspect of her reality. There was beauty behind observing how the family listened, respected, and shared in each other's world.

"Everyone in my head is a different color and I roll them around and sometimes they can change. So, the colored balls, they each have an individual color based on what I think about you, okay? And the color that I think about my mom is blue because she's like blue, like the sky. Then there's my friend, Pi, who is obviously red . . . actually, I would say she is red because she is a reddish-blueish person. And then, Christine would be purple, because she is purple for her whole life and then you're being green because you have a lot of stuff that makes you green as well. Red is like fierce and ready to go. Blue is just like . . . it's kind of peaceful."

I glanced at Terry. "That's Rosie. She groups people like that, too. Always striving for order."

"And then purple, of course, I think would be in the dauntless category. Green would be selfless like abnegation. Just kind." Terry added a natural acknowledgment, "Oh, that's nice. That's so nice."

Rosie flapped her hands, then rolled her head forward and backward and around as if she were on a Tilt-A-Whirl. "The balls also move, like this," she explained. "And then there are some colors that you don't want at all. But, if someone turns their life around, then their color will start changing and the color will begin to fade. You can also go to those colors if you are nice in the beginning and then you stop being nice. The balls are all the same size and they look like shiny bowling balls and even move into rows."

She waved her hand over an invisible area in front of her face. "Another thing is that my brain also projects like a movie screen. Whether I'm watching a movie or not, my brain projects movies.

It'll project like this scene right here and then I would put it there for the present, so it's like the present. So, it is very visual. It's very visual."

As I inquired further, it gradually became more apparent that Rosie was describing that she keeps track of her memories like films that she can replay for reference. Terry might ask Rosie about a fact or an event that happened in third grade, and Rosie will mentally turn to *Rosie Movie 3,* look for the scene, and then can describe it with superb detail while reviewing the film of what happened.

"When I see people," Rosie explained, "I also use my movie screen, too, and I reflect them onto the movie screen and, I know this sounds a little strange . . ."

I was struck when she mentioned this because it was the first time she had demonstrated some *theory of mind,* her own estimation of what another person might be thinking, which stands as a representation of how far her family had helped her develop her social cognitive skills.

"The people I see will be reflected onto the movie screen. And that person . . . They'll reflect off the movie screen and sometimes . . . Dah dah daaah! Music will happen." It was the kind of plot-twisting cinematic score that moves down the music scale and introduces a villain or lays out a new layer of suspense. "Sometimes it will be happy music when a nice person comes out, which is most of the time."

Aptitude for visual thinking or novel sensory perception is relatively common in autistic spectrum disorders. Sometimes the diversity of sensory integration and processing can manifest as synesthetic association or simply associated percepts between broad categories. Daniel Tammet, for example, carries the moniker of the "Brain Man" for his superhuman ability to quickly learn languages and memorize large swaths of information. He is the

European record holder for being able to memorize and recite *pi* to the 22,514th digit over the course of five hours. Like Rosie, he is on the autism spectrum and had epilepsy as a child. His world, too, is filled with bright and distinct qualia to the point where he changed his birth name, Daniel Paul Corney, to Daniel Tammet because the associations of his given name simply "didn't fit" with his perception of himself. At the core of Tammet's mnemonist ability is his tremendous synesthetic library in which every positive integer has a unique shape, color, texture, and emotional feel all the way up through the number 10,000. Cambridge University's Dr. Simon Baron-Cohen surveyed 164 adults with autism and found that about 20 percent of them had synesthesia, a rate of synesthesia three times higher than people without autism. Both autism and synesthesia are hypothesized to involve abnormal neural connectivity and thus may have some biological linkages. At the Wellcome Trust Centre for Human Genetics at the University of Oxford, Dr. Anthony Monaco, working with Baron-Cohen, has identified at least four genetic regions associated with an increased likelihood of having synesthesia, particularly chromosome 2q24, which is also tied closely to autism.

Making use of these connections and associations to help build more relatable perceptions and interpretations of the world help provide Rosie and others like her some sense of meaning in an otherwise alien world. Rosie's internal constructions of the external world assuage many of her anxieties, which gives her a sturdy rudder for steering her thoughts, though she has difficulty remembering the faces of people she meets. To help her recognize some people, she describes using a "voice recorder" in the back of her head. When she hears a person's voice, she goes to the voice recorder, turns it on, and compares what she's hearing with the recorded catalog of voices that she keeps so she can figure out who is talking.

"I picture the sound of music in my brain. A and B on the piano are white. C4, C5, and C8 are a blue sound. They're peaceful. Music calms my brain down. It makes me feel very good. It decreases my seizures, too." Terry brought her left hand up to lightly touch her lower lip and paused. "We had finally started reading . . . and then the seizures hit real bad." Unknown to Terry, though, the seizures had already been occurring for most of Rosie's early life. While Rosie was playing, she might suddenly stop all movement and, after a few seconds, resume exactly where she left off. Terry assumed that she was just distracted. What may appear to be a brief and sudden distraction to a bystander, however, may actually be an absence spell, previously referred to as petit mal seizures, or perhaps more aptly described as brief nonconvulsive generalized seizures.

In medical school, I recall seeing the video of a small boy in a blue sweater with EEG electrodes pasted all over his head, tendril-like wires projecting backward as if he were the illegitimate son of the alien in the sci-fi flick *Predators*. He sat on his mother's lap while holding a small red pinwheel in front of his face. The disembodied voice of the EEG technologist behind the camera cued in a cheerful tone, "Go ahead and blow! Like you're blowing out a birthday candle!" The child began to blow on the pinwheel. "Keep going! Keep going!" After blowing out eight imaginary birthday candles, his breaths suddenly stopped. The pinwheel continued spinning. The little boy appeared empty, lifeless—as if instantly transformed into a wooden Pinocchio. On the right side of a split screen, the EEG demonstrated rows of synchronous squiggles deflecting up and down sharply followed by a slower deflection that appeared like frayed mittens lined vertically side by side, a telltale pattern of "generalized 3 Hertz spike-and-slow-wave epileptic discharges." For a few seconds during the abnormal epileptic activity, the boy remained lifeless, stuck. But then, once the epileptic seizure stalled, the boy blinked, yawned, then resumed blowing on his pinwheel. Absence seizures like this can occur in almost any child, but children with autism are at a much higher risk. Fortunately for many children with epilepsy, seizures can be treated easily with well-proven medications. The absence seizures themselves, though clearly disturbing, are usually harmless. Sometime during puberty, childhood seizures often spontaneously resolve naturally, and a young adult can eventually get off his or her antiepileptic medication.

Rosie wasn't so lucky. In kindergarten she progressed from traditional absence spells to focal seizures, which may or may not impair awareness. As the name implies, focal impaired awareness seizures disrupt a person's cognitive abilities, which causes them to experience a lapse in consciousness and memory or a dramatic change in behavior, like suddenly smacking their lips, picking at their shirt, or becoming violent. Conversely, a focal aware seizure leaves the person completely conscious but usually with a rhythmic shaking. The older nomenclature for a focal seizure with impaired awareness included *complex* as a modifier for the seizure. Thus, a *complex partial* seizure in the traditional epilepsy parlance describes a focal seizure with impaired awareness where the seizure disrupts and commandeers the electrical rhythm of only a part of the brain but still alters consciousness. Multiple EEGs showed that Rosie's seizures occurred in her right and left temporal lobes, the parts of the brain most responsible for emotional processing and the storing and retrieving of memories.

Epileptic discharges can be brief or prolonged and can spread like a brushfire across parts of the brain. The most dangerous form of epilepsy, *status epilepticus,* is a seizure or cluster of seizures that lasts longer than five minutes without the person returning to their usual level of consciousness between or after seizures. In status epilepticus, immediate action is required by a medical team in order to stop the prolonged seizure before neurons in the brain start to show signs of cellular damage. This is why the most important action when witnessing a seizure is keeping track of time and getting medical help, not placing objects in mouths or attempting to physically stop the convulsions.

Outside of the dangerous condition of status epilepticus, epilepsy is typically treated more like a chronic medical condition. Treatment starts by initiating an antiepileptic, or antiseizure, medication. The first antiepileptic medication that is prescribed usually helps improve how often seizures happen but less than half of the time will not be able to completely prevent seizures from recurring. The next step involves increasing the dose of the antiepileptic, which by the nature of how these medications work can often dampen the brain's electrical activity. Once a high dose is reached, or if the side effects of the medication start to interfere with quality of life, the original antiepileptic is switched over to the next-best agent. At this point most patients will finally have a stable medication that works, but a smaller fraction will continue to have seizures. When multiple solid attempts at controlling seizures with different medications have failed, the epilepsy becomes *refractory,* difficult to control. For these seizures as well as those that have a very specific cause such as an abnormal spot of tissue in the brain, the only true "cure" involves surgery, though invasive procedures can often pose more risk than reward. Some patients, like Rosie, are not candidates for surgery because the risks of residual cognitive deficits are likely while the chances of a cure from the procedure are remote.

In lieu of surgery Rosie tried at least seven different antiepileptics. Some medications helped improve the seizures but, at the same time, they made Rosie irritable, hijacking her mood with more fervor than a seizure. Other medications, while more tolerable, did not have much effect. At five years old, Rosie was started on a powerful medication called lamotrigine, initially used to treat focal seizures and, later, as a mood stabilizer for patients with bipolar disorder. While effective in stabilizing seizures, lamotrigine can also lead to a life-threatening autoimmune reaction. Mounting an attack against themselves, immune cells kill native cells on the skin and mucous membranes throughout the body. The tissues come apart like paint scaling off of an old wooden barn. For Rosie, doctors started her on low doses of lamotrigine, slowly, over weeks, until they would find a dose that reduced seizures. Rosie started refusing meals soon after Terry began to increase the dose of lamotrigine as prescribed. Most of the time she still couldn't speak. When Rosie started having difficulty swallowing, Terry looked inside Rosie's mouth. A few sores had begun to form on the inside of her mouth. "We took her to the pediatrician who found that she also had blisters in her feminine areas. That was when we first learned about Stevens-Johnson syndrome."

The first patient I ever saw with Stevens-Johnson syndrome was a fourteen-year-old girl in the pediatric intensive care unit. She was smothered in weeping lesions. Her blistered mucosa was bloody and in some places, purulent, including the insides of her mouth, her lips, and her vulva. Her eyes were the color of raspberries, trapped behind swollen eyelids, oozing the copious gel applied to provide moisture. The weeping blisters flaked off layers of dead cells like thin wet leaves. The mirrored sensation felt like the skin on my body had been corroded and chewed away in patches. I felt the breathing tube in my own throat, the sensation of cloth bandage tape tightly across my mouth. The sweat-covered strands of the girl's wavy dirty blonde hair smeared across her forehead echoed a minor discomfort that paled in comparison to the moist crinkling across her entire body, seething raw with anger—bloated, betrayed, obtunded, mercilessly disintegrating.

"It was so painful," Rosie said, recalling her experience in the ICU with early-stage Stevens-Johnson syndrome. "I couldn't say anything, but it hurt. I just hoped for the best and stayed positive."

The lamotrigine was discontinued soon enough that Rosie only suffered a partial reaction. In some families dodging the worst of this macabre reaction would have served as a clear signal to abandon hope of normalcy. For the Dohertys, though, it was received as a reinforcing call to arms. "I'm so happy my mom didn't give up on helping me talk. She kept going. She was like, 'I'm going to get this girl to talk and this is how I'm going to do it.' It was a lot of hard work actually, but by the time I was seven I was pronouncing words pretty much fine. I didn't talk in full sentences, but at least I was able to say things like 'Love Rosie.'"

By then, Rosie could pronounce enough words to practice talking through echolalia, repeating other people's words. "I was practicing the sounds. I would repeat back so I could learn the sounds because if I didn't, then I wasn't going to talk. It took a lot of work to be able to repeat back all the sounds they were saying. By repeating I was able to say 'Rosie likes orange juicy,' which was something I came up with on my own." Rosie continued to make progress and had finally transitioned from non-verbal to verbal. But she experienced an unexpected setback.

Rosie was started on a new antiepileptic called Topamax, also known as topiramate. The sedation from the medication was visible enough in Rosie that the family picked up the nicknames for the medication, "dope-a-max" and "stupamax." Making matters worse, the topiramate was also ineffective for controlling Rosie's seizures. It had become clear when she had a seizure that lasted longer than usual. One day Rosie stopped talking midsentence, prompting Terry to keep her eye on the clock. One minute . . . two minutes . . . This was unusual because Rosie's seizures typically would, at maximum, only last about three minutes . . . The second hand on the wall clock continued to move and Rosie remained frozen. Four minutes. Terry picked up the phone and called an ambulance. Terry reached for her purse to administer the emergency rectal diazepam, which is often used to abort seizures in children when they are unable to swallow. She tore off the clear plastic wrapping of the rectal syringe, turned Rosie over with one hand, and pushed in the plunger, hoping it would break her seizure. Later, in the emergency room, Rosie had begun to recover, but because she was not completely back to her usual self after close to thirty minutes, doctors ordered an emergency dose of intravenous phenytoin, an antiepileptic, which put Rosie again at risk of developing Stevens-Johnson.

The next afternoon, a blister appeared on the roof of her mouth, followed by a second, then a third. Within a day, it became clear she was experiencing a mucocutaneous reaction. Rosie's pain intensified and was barely able to utter in her weak raspy voice, "No orange juicy for Rosie."

She developed a painful desquamatinous rash over 5 percent of her body. Then 10 percent, then 15. Likely because she had only been exposed to a single dose of the medication, the progression slowed and stopped at 20 percent. Her body was spared, but the setback narrowed further what few options for antiepileptics she had. After starting a medication known as levetiracetam, Rosie's seizures began to wane. Now in the fifth grade, the nine-year-old Rose was finally able to read through the help of phonics. She made occasional errors such as pronouncing "it" as "eht," but at the very least she was on her way from going beyond "behind" in school to "just a little behind."

Eventually, Rosie entered puberty, which brought with it its attendant hormonal swings. Some families consider a first menstruation as a matter of celebration and a first rite into womanhood. However, the increase in estrogen after menarche lowered her brain's intrinsic ability to prevent abnormal electrical activity. Which is right around the time "the seizures hit real bad." Menarche marked Rosie's first generalized tonic-clonic seizure. Having witnessed many seizures as a neurologist, the aspect that seems most disquieting is how quickly a generalized tonic-clonic seizure can progress through the body. First, there is an unexpected, short-lived silence. The muscles of the entire body begin to contract in rhythm with electric discharges in the brain that run through the motor cortex. The lack of control in the body, as it is possessed by an unseen force, causes the arms to pulse and push outward in the seizure's clonic phase as if attempting to fight off the invasion, followed by a full loss of control, every muscle contracted and stiff—a rigid tachypneic corpse with eyes fixed open, glazed, distant, and lifeless. Finally, the entire body shakes violently, a temblor tearing through with vile accuracy, an epileptic possession.

Rosie was having generalized tonic-clonic seizures up to three to four times in a week, a small natural disaster devastating her world almost every other day. With each epileptic conflagration in the brain, the residual electrical activity resembles smoldering ashes, a fire ready to rekindle. Terry lost count of how many smaller seizures Rosie was having daily.

Even on four times the usual dose of levetiracetam, Rosie's seizures were occurring so frequently that her brain began to lose the ability to learn and process information. With each seizure, her ability to read was slowly chipped away. But for Rosie, the most disturbing parts of the onslaught of seizures were the "black holes" in her memory. Rosie had taken comfort in remembering minute details of the world around her. It soothed her worries about uncertainty because her affinity for autobiographical memory made her daily life seem more recognizable. Disarmed, Rosie gradually withdrew.

Terry was desperate for a solution. A friend of hers mentioned that her son had been seeing a pediatric epileptologist named Dr. Elizabeth Thiele, who started him on a special ketogenic diet that was working for them. Terry was willing to try just about any option but hesitated. It sounded too *New Age,* too fringe. When Terry asked their epileptologist about it, he shrugged it off, suggesting instead phenytoin, despite the high likelihood of another traumatizing skin reaction. With a firm nudge from Christine, Terry picked up her phone and made an appointment to see Dr. Thiele.

Several days later, they found themselves in a new waiting room, unsure of what to expect. Rosie sat in her chair looking at the floor, cheerfully swinging her legs. She suddenly stopped moving. A seizure took her brain hostage. Her head tilted back softly, and she fixed her gaze into the abyss of epileptic emptiness. She urinated in her seat. A few seconds later, when the seizure's grip over her mind loosened, Rosie returned and Terry began to gather her belongings to take Rosie to the nearest restroom. Before she could, however, a voice called out, "Hi, I'm Dr. Thiele. Why don't y'all come back?"

Rosie beamed at the memory of meeting Dr. Thiele, "She was so smart. Boom, boom, boom! She knew everything by looking at me." Rosie snapped her fingers high in the air. "She's *this* smart."

In a calm and welcoming tone Dr. Thiele walked Terry and Rosie through why Rosie was having seizures, why she had difficulty controlling them, then discussed their options going forward. "Dr. Thiele didn't want to rely exclusively on the diet just yet," Terry told me, "because her seizures were so out of control. She started her on rufinamide and the absences"—she raised her hand in the air and snapped her fingers—"the absences were totally controlled by the rufinamide." The addition of the rufinamide, combined with a Canadian-manufactured drug clobazam and medroxyprogesterone (a "depo shot") to counteract the neuronal excitability and seizure-activating effect of estrogen, finally gave Rosie a reprieve. They were able to reliably gain better control over the seizures. After a few months of minor medication adjustments, they were finally ready to talk about *the diet,* which imitates the effects of starvation.

After a starving body exhausts its supply of glucose and other carbohydrates, stored fats become the primary source of energy. Metabolizing fats leads to the creation of ketones. In the ketogenic diet, caloric intake comes mostly in the form of fats with a standard recommended daily allowance of protein and as few carbohydrates as possible. Like several common medications with unclear mechanisms of action, the antiepileptic effect of ketogenesis remains a matter of speculation, even after a century of research. The ketogenic diet, through chronic elevations in ketone bodies, likely acts in three ways: modifying the body's mitochondrial function; increasing production of inhibitory neurotransmitters; helping neurons to become less sensitive to electrical triggers.

Back when I was still a neurology resident rotating through pediatrics, I learned about the idea of diet as therapy, food as proto-drug. To understand the diet more intimately, I decided to go on a ketogenic diet for a month. The experience was surprisingly valuable, particularly in the more technical aspects of my clinical practice. One night I was paged to come downstairs and evaluate a three-year-old girl with a hereditary form of epilepsy in the Pediatric Emergency Department. As I pulled back the curtain, a feeling of exhaustion washed over my body. Before me were a young mother and father lying in the stretcher with faces covered in a mix of grief, fatigue, and desperation. Their daughter had Angelman syndrome, a genetic neurodevelopmental disorder where children demonstrate a combination of intellectual disability, seizures, and a constant almost puppet-like demeanor of frivolity and laughter. This girl, though, had been crying for about three days straight. Her parents related how she had been doing so well with seizure control while she was on her most recent medication, in combination with ketogenic therapy. Now, though, she lay in her father's lap, wailing and shrieking, kicking and writhing. The emergency medicine resident standing at the doorway folded his hands across his gray sleeveless fleece. "She doesn't look like she's seizing and there's nothing in her history that makes it sound like she has a cold or anything."

I wasn't so sure. When someone is uncomfortable or in a state of obvious, agitated distress, the agitation *is* the symptom. I looked at the parents and felt their concern for their child sink further into my jaw and shoulders. I paused for a moment and recalled some of my own experiences while in a ketogenic state. I turned to the emergency medicine resident "Let's check a bicarb," I told him, "and a beta-hydroxybutyrate level in her, too."

Because beta-hydroxybutyrate is the first ketone produced in starvation, it is an objective biological marker for ketogenic dietitians to determine when a child on ketogenic therapy has finally entered a state of ketosis. It also helps give a sense of how closely a family has been able to follow the strict dietary regimen. An intermediate of carbonic acid, bicarbonate also plays a key role in the body's natural ability to regulate acidity in the blood. Though the body's acidity is expected to increase while in ketosis, it can become too high in a state of infection or physiologic stress, also known as ketoacidosis. The acidity can interfere with the body's ability to metabolize and involuntarily upregulate a person's respiratory rate to eliminate excess acidity through the lungs in the form of carbon dioxide. This same danger is common in diabetics who lose the ability to remove and use glucose from their bloodstream and can swing into a state called diabetic ketoacidosis, which needs to be treated emergently with aggressive volumes of intravenous fluids to assist the body in buffering the acidity and treating severe dehydration.

The girl's test results showed that the bicarbonate was abnormally low while the beta-hydroxybutyrate level was dangerously elevated. We discovered that she recently had a short bout of soft stools that, after additional discussion, was determined most likely to be diarrhea from a stomach flu. Combined with her refusing of food from her upset stomach, she had lost enough fluids to tip her over into ketoacidosis. Heidi, the hospital's ketogenic dietitian, was able to follow up with her inpatient team with recommendations on how to correct the acidosis and, within a day, the family was able to return back home—back to their more usual state of calmness punctuated with frequent moments of angelic laughter, all while keeping her seizures at bay.

Rosie's new diet introduced more uncertainty in her world. The uncertainty led to anxiety, which led to fidgeting and irritability. It was a challenge for the family as well—having to deny Rosie's requests and becoming even more unsure about what she needed. "I remember this one time," Elizabeth said, tucking a rogue strand of red hair behind her right ear. "We were in the grocery store and Rosie had just started the diet. She looked at the back of the labels of everything that went into the shopping cart. My father put a box of popsicles in the cart and Rosie picked it up and yelled, 'There's only six carbs in these popsicles! I think I can have them!'"

Terry covered her eyes imagining herself back in the freezer aisle. "And the people around us looked at me like, 'How could you put this skinny little ten-year-old girl on a *diet*?'"

I felt Rosie rubbing her forehead vigorously with her left palm and shove her bangs backward. "When people hear about the diet, they can get weirded out and they feel so bad for me. They almost treat me like a baby with so much pity and they'll tell me that it's 'devastating.' Like it's really devastating that I can't have a piece of cake."

Christine, revealing the green impish 6 tucked within her prominent blue hues of 4 and speckling of turquoise 7s, added, "Some people just feel awful for us like we're being deprived of something important."

"Yeah, we're being deprived of seizures. That just breaks my heart, right?" Terry let out a mirthful laugh along with the rest of her girls.

Rosie's seizures improved substantially on ketogenic therapy and a few more medication adjustments.

"Even my horrible belly aches that I had and that a lot of people with autism have . . . The minute we were on the diet, it was like you flicked a switch." Rosie snapped her fingers like a whip. "Gone. Gone! Heidi and Dr. Thiele are my heroes. They helped me out and now I'm living like this, which is great!"

Once the hurricane of seizures settled, the Doherty tribe regrouped. Steadily, Rosie relearned and reclaimed the functioning she had lost.

Remembering how music often calms down Rosie's brain, I asked her if music also helped decrease the frequency of her seizures.

"Okay, so I have a very specific-looking aura. I like to call it 'the demon.' It really, it really scares a lot of people if I call it the demon, so I call it 'the red thing' now. It's red, okay? And it has these black, and then it has these black little horns on it and then it has these . . . then it has these eyes." She widened her eyes and lowered her chin toward her chest. "Then it has yellow circles like the one in van Gogh's *Starry Night.* Then, if it gets worse, it starts getting a green circle and then it goes into rainbow colors and then that's when I go down, when it gets rainbow colors. You don't want it to get rainbow colors . . . ever."

Elizabeth ended Rosie's description with a warning. "Don't get the rainbow."

"When I know I'm feeling nervous that day, I might start to get sweaty hands or I might start pacing or I might see some of my aura, I'll listen to music. When I listen to Billy Joel or Bob Dylan or Bob Marley, I'm thinking about that music and I'm not thinking about like a hundred other things, like 'Oh my god, am I going to have a seizure?' When I relax, I feel Bob Dylan's music. Or I'll play the piano. I'll play the piano in my head or I'll play it in real life. I'll sing and rock, then I start calming down. And sometimes when I'm really nervous, like when I'm about to go to a huge event, I imagine myself that I'm there . . . and then, I hear the keys of the piano. I hear whatever song I want to play and I will imagine that they are singing while I continue playing on the piano. Listening to music is where I feel the *most* most normal. It comes from my heart and I don't feel like I'm in the outside world. I feel like a normal person. So normal. I don't feel like an autistic person. I don't feel like an epileptic. I feel like a person who's in that same exact person who's singing whatever music I'm listening to. I can feel their passion. I can feel what they're feeling inside. Billy Joel is very good at this. He can paint the picture with all of his words and, when I sing his words, I feel what Billy Joel is feeling inside. I can feel the notes. That's where I feel the most normal and I feel it all around me and everything else just becomes like fog."

Through her heightened senses, Rosie has learned to tap into metacognition, or mindfulness. Attending to the nuances of any of the senses creates an opportunity to notice and become an observer of one's own emotions and thoughts. Sounds, textures, or even the sensation of our muscles being put to work or the positioning of our own bodies can dilate time. It allows us an opportunity to play a more active role in what thoughts we entertain, to acknowledge the visceral sensation an emotion has flushed throughout our body so that we can decide for ourselves to engage in whatever activity it takes to elicit the relaxation response. In this state, a person can lower his or her heart rate, respiratory rate, blood pressure, or increase the activity of the "rest-and-digest" part of the nervous system.

"When thoughts enter my head that something bad might happen," Rosie told me, "I just get rid of them. I start by singing songs or I might imagine Bob Dylan or Billy Joel holding up a sign saying, 'Don't be nervous, Rosie. You're going to be fine.' Even if my aura is over there saying,"—and here Rosie makes a snarling, demonic-sounding voice—" 'Wanna have a seizure, Rosie?'

"When I pet our dog Mandy or just think about petting Mandy, that helps, too. I do pet the dog because the dog can tell when I'm nervous and she will sit next to me and I can feel the fur and it helps. When thoughts enter, I sing, pet the dog, or I run. I'll run, run, and run around the driveway. I'll run and sing at the same time if I have to."

Terry added, "It's excellent therapy."

While we may not all be yogis on a mountainside who can elicit this relaxation response with a mere thought, we can at least breathe. Reflecting on the nuances of our breath or acknowledging our sensory experience is as much a contemplative practice as meditation or yoga. This is the built-in beauty of mindfulness. It can be just as easy to be mindful as it is to be mindless. We have all the hardware and software we need to be able to tap into a contemplative practice that will allow us a greater locus of control over our mind and the ability to select our response to whatever occurs in our external or internal experience.

"The fact that I have seizures can't be undone. There's nothing I can do about that. It's just the way God made me and I have to accept the way God made me. But, I can *live* with it. It can be difficult sometimes, but I have to think the best for me. 'I'm not going to have a seizure today.' And sometimes I do, but most of the time I don't. I just think positive enough. 'I'm not going to have a seizure today!' Even if I do, I still think the best for me."

I physically felt the expressions soften on Terry, Christine, and Elizabeth's faces. Rosie was sharing the story that she had rewritten in her own words, her own language, huddled closely with her tribe when she needed them most, through seizures, fear, and the inevitable catastrophes of our internal and external worlds that are inseparable from living.

"My biggest trigger for seizures is stress and that taught me that your worst enemy is pitying yourself too much. 'Oh, poor me, blah blah blah . . .' If you just keep pitying yourself, nothing changes. Nothing in your life is going to change because you keep pitying yourself with the same old story. When you stop doing that, you give yourself a chance to feel better and change what you can so you can have a better life. I know I had obstacles, but I told myself I have to keep living my best no matter what happens because that's just the way life is."

Terry looked to Rosie, crying. "Where did that come from, Rosie?"

"From you. You taught me how not to pity myself. 'You keep going. You be strong.' My mother taught me to do that."

Rosie's remarkable resilience and compassion for herself was not encoded in her DNA nor did it suddenly appear one morning. Her resilience and compassion for herself and, remarkably, for others rests on years of telling and retelling her story, reflecting on the significance of the joys and struggles that have visited her and her tribe.

Rosie may score low on conventional measures of empathy. But with her tribe, she has nurtured a tremendous capacity for compassion by experiencing their narrative. Empathy may be the seed of compassion, but we must lovingly tend to this seed with devotion, with each other. While I had been relying too heavily on my reflexive mirror-touch synesthesia, Rosie opened herself up to others while remaining firmly rooted in her tribe, to share the joy and suffering cultivated in her.

CHAPTER 8

A Running Start

ONE MORNING IN THE EMERGENCY ROOM, WHEN I WAS STILL a first-year neurology resident, I started off with a handful of consults. The steady flow showed no sign of slowing. While wheeling an acute stroke patient to the CT scanner, I felt the look of panic on a nurse who was on the phone. The thin edges of her eyelids were stretched as if they were about to tuck backward behind her wide bulging eyes. Her mouth hung open as the muscles in her mandible slacked almost in disbelief, and she stared straight ahead, her eyes fixed on some unseen mental image. For a few tense moments, she stood quietly in place, listening to the caller on the other end of the receiver, then yelled, "A bomb! A bomb went off at Copley at the marathon finish line!"

I reviewed a list of possible scenarios and potential solutions in my mind as we brought the patient onto the bed of the scanner. I stepped into the technologist booth. Time for action. I got on the phone with my senior resident to give him a briefing on the situation. "I'm going to need backup. At least until all of the neurological patients are out of the emergency room to make room for the incoming trauma cases." He rushed down, assigning available residents and medical students to spread out and take point on each of the consults. Together, in less than fifteen minutes, we were able to find hospital beds for each of the consults, including the acute stroke consult we were in the middle of running, and yielded the floor to the emergency room and trauma teams.

As I passed by a long line of stretchers wheeling in marathoners, the soot, cuts, and charred flesh brought me back to the sights, the smells, and the feelings I had experienced during my car accident. I had a store of personal memories from the accident, all linked to emotional reflexes, the urgency of impending danger and the need to either face it down or flee. The heightened salience made each physical sensation harder to ignore. They felt real. Was I experiencing their trauma vicariously or personally? The distinction between these two experiences of the self and other seemed illusory—the words *personal* and *vicarious* had lost their utility years ago. The story that I was composing for myself in the middle of this trauma was all that would matter, today and every day after.

But I couldn't help wonder which themes the victims were weaving through their stories. Perhaps they would pen their story with the ink of pain. Or maybe they would tell their stories with fear, anger, spite, futility, distrust, and betrayal scrawled along the margins. Though months later, as the people hurt in the bombing recovered, these marathoners instead were telling a different story. These were stories of human resilience, of bouncing back stronger than before, of slow laborious persistence without the trumpeting of fame and glory.

"We tell ourselves stories in order to live," Joan Didion reminds us, and our ability to create meaning out of personal tragedy is critical to our existence, to rediscover resiliency while coping with adversity. Throughout my career, I have been reminded of this time and time again by my patients and in my research. Reclaiming personal narratives in some way appears to be a key protective psychological and biological factor in recovery, particularly in victims of traumatic events, like people who have experienced a stroke. Our conditions, our circumstances, are too often out of our control. But the stories we tell about these circumstances are ours and ours alone, allowing us to choose the way we live.

I first met Teri at the annual International Stroke Conference. A critical care nurse and one of the most sought-after coordinators of stroke programs in the Midwest, Teri had recently survived a stroke, an irony she acknowledged right away. "Knowing *exactly* what I was going through *while* I was going through it was more than a little distressing."

She had just arrived from Minnesota, dressed for overnight travel in a Jonagold-red velour tracksuit. A distinct joy flickered from her smile, her slightly drooping face, like light reflecting off the ripples of a lake. To my delight, she was a pronounced combination of speckled Russian-blue 4s, a bold maternal red 2, a prominent Midwestern yellow 8, and a shading of white 0s, which complemented her short golden hair perfectly. Whenever she smiled or laughed, which was often, the creases around my eyes crinkled in quiet recognition of her candor and openness.

I liked her right away. Though everyone liked Teri right away. Everyone at the conference knew about Teri. She was a kind of cult hero. Not only was she the stroke program coordinator who had a stroke, she was also the woman who ran a marathon four weeks after suffering a stroke.

When I sat down with Teri, I had previously ended my first year of neurology and was well into my initial stint as a senior resident, placed on the demanding stroke consult service. Because this service was notorious for being one of the most challenging rotations of the second year, I considered my assignment to the position of senior before the end of my first year an honor but also a source of apprehension about having that much responsibility bestowed on me so soon. After a few weeks as a senior, taking the time to teach medical students, medicine residents, and new neurology residents, it became incontrovertible—I had in fact been learning and, to my disbelief, had plenty of knowledge to dole out as well. As a result, during my last year of residency, I took a terrifying risk. I refused to declare a narrow interest in any of the existing neurological specialties. Instead, I declared a general interest in "brain health." Having seen the end of a wide array of chronic devastating neurologic conditions, I wanted to dedicate my contribution in the field to helping guide people toward what I had found in my experiences—in my sensations and in my interactions with them—that matters most: living a fulfilling life. It meant, in one extreme, going back upstream early enough in the history of disease to prevent it from occurring in the first place. While in the other extreme, it meant wading around midstream to find new, even nondrug, methods to help patients preserve the function and quality of life that they had, especially in active neurological diseases with the potential or expectation to progress.

Much of "brain health" in the mainstream, however, had become a thick quagmire of recommendations with varying degrees of evidence for their support. I fed my desire to contribute as a clinician-scientist by pursuing specialized training in the skills of a behavioral neurologist and neuropsychiatrist. In this way, I told myself, I could look at the brain in the entirety of its functioning, especially thinking and feeling. At the same time I could also develop the research expertise to gain a better understanding of what existing evidence provided: a balanced approach to promoting health and wellness, which strives for a happy medium between what scientists describe as the "soft stuff" and what nonscientists describe as the "cold, hard, uncaring sciences."

I wanted to bridge the two.

Stroke is an eventful injury to the brain with an observable change in function followed by a period of heightened brain recovery. As such, stroke seemed to be a natural starting point to begin building the bridge between these two worlds.

Teri was my unexpected keystone.

As Teri and I sat down together, I felt her right facial muscles tighten slightly on the left side of my face. A trace of a cramp, a minor spasm or twitch, crossed my left cheekbone. The sensations were fainter, more distant, along the lower half of my right face, which mirrored the weakened muscles on the left side of Teri's face. All of these impressions were consistent with Teri's physical condition following her stroke. They weren't necessarily holding my attention. Not really. What was holding my attention was the odd sensation of movement in my left arm, even though it rested motionless on my tabletop, and the soft whispering around my hairline. Why were these sensations lingering, I wondered. Why were they so pronounced? What were they trying to tell me about Teri? And then it hit me. Every so often, as Teri started in on her story, she fidgeted with her bangs using her right arm, tidying up the slightly darker roots near her part. She was self-conscious about her appearance.

Teri suffered her stroke on Memorial Day 2013. Despite a torrential downpour, earlier that morning Teri finished a six-mile run as part of her training for the 26.2 Grandma's Marathon, which ran along the North Shore of Duluth. Because it would be her first marathon, she meticulously scheduled and outlined her training regimen, determined no matter the setback to transition from *runner* to *marathoner.* After a quick shower, she drove off to visit a local elementary school to teach fifth graders how to spot signs of a stroke. With her that morning was her sixteen-year-old son, Parker, who often helped Teri with her stroke education programs. He needed the community service hours to graduate from high school, and Teri enjoyed getting to spend a few extra hours a week with her son. They stopped at Starbucks, a ritual cultivated together over the past year, and on their way back to the car, Parker offered to drive. Preferring to be the one in control, Teri declined with a maternal smile.

With a hot Venti cup of Blonde Roast coffee in her left hand, Teri steered the car down the streets of their tiny Midwestern town, easing peacefully into her morning routine. She approached a flashing stoplight, stepped lightly on her brake. Pause. She gently raised her foot off the brake and rolled forward. The light pulsed red. Gone. Body. Arm. She had lost all feeling in her left arm. Her coffee cup started to slip out of her hand. She turned to Parker to ask him to take the coffee. Silence. Her mouth was devoid of words. She struggled, her tongue and diaphragm pushing desperately against the silence. Numbness slithered up her neck to her cheekbone. She felt as if she were being swallowed alive.

Her first thought was: *I can no longer protect my child.*

The second: *I'll never be able to say "I love you" again.*

And, when the gravity of the situation firmly took hold, the third: *I am going to die.*

In a daze, nearly overcome with fear, Teri pulled the car over. Saliva dribbled out of the corners of her mouth. Parker's eyes widened, "Mom, do I need to call 911?" Teri shook her head no. But Parker knew better. "Mom. You're having a stroke." Teri pointed at the dashboard clock. 10:30 A.M. The time seared into Parker's consciousness. 10:30 A.M., a crucial detail. 10:30 A.M., the time his mother, in the parlance of her stroke education programs, was *last known well.* Though Teri could barely raise her left arm, she still had use of her left leg, which she relied on to thrust her body over and switch seats with Parker.

Seven minutes later, they were in the emergency room—an earthly miracle. Teri had stroked a half-mile from a primary stroke care center, literally the only place in Duluth that could help her at exactly the right time she needed it. There is no waiting in strokes. Time is brain; brain is time. Teri knew this all too well. She lay upright on a stretcher. A physician shouted questions at her while nurses darted around her. Amidst the chaos, Teri focused on the ticking clock hanging in the left corner of the emergency room. Her mind ran through the same statistics she had rattled off time and time again to women her age during her stroke education programs.

A stroke occurs every forty seconds.

One in six people experience a stroke.

Stroke is the third leading cause of death in women after heart attack and cancer.

Stroke is the leading cause of long-term disability.

With every minute, Teri knew, she was losing two million brain cells. With every minute, she remembered, her brain aged by a month.

The second hand on the clock was merciless.

If the medical team did not successfully return the normal flow of blood to her brain, she could die of aspiration pneumonia in thirty days. She might end up with a plastic feeding tube surgically implanted through a hole bored into her abdomen that connected her stomach to a bag of liquid meals.

She wondered if she were going to be one of the 20 percent of stroke patients with communication issues who contemplate suicide.

She was fixated on the clock to her left while the medical team wondered if her gaze was indicative of the severity of her stroke or a seizure or a residual Todd's paralysis, a focal weakness in a part of the body after a seizure. Teri followed the medical team's commands to raise her legs and arms, to squeeze their hands, so they knew that she was still *in there.*

She looked at the clock again. Two million more brain cells.

This was terror, and she was fanning its flames.

She had to stop herself. She had to turn off her stroke nursing coordinator brain. Be a patient, she told herself. But the clock continued to taunt her. She closed her eyes. The best she could do at that moment was to think about her husband and her son.

And pray.

STROKE WAS FIRST RECOGNIZED AS A MEDICAL CONDITION BY Hippocrates and was described as *apoplexy* (meaning "struck down by violence"). Colloquially, the word "stroke" refers to two primary types of injury to the central nervous system involving blood vessels, hemorrhage and ischemia. Swiss pathologist and pharmacologist Johann Jakob Wepfer was the first physician to suggest people who suddenly fell paralyzed or mute by an invisible force, the "stroke of God's hand," were in truth victims of cerebrovascular bleeding or blockage. Hemorrhages in the brain can be due to a list of potential causes that include an aneurysm, the rupture of a cerebral blood vessel, or high blood pressure, which can cause the thin wall of a small blood vessel to break open and bleed. Each cause of cerebral hemorrhage has a specific treatment though all abide by the same basic medical principle: stop the bleeding immediately. Bleeding in or around the brain is the immediate concern; once that happens, the primary damage has already been done. In select instances, such as in a subdural or epidural hemorrhage where blood creates a layer on top of the brain, practiced brain surgeons can remove the blood. But in most cases of subarachnoid or intraparenchymal hemorrhages, removing the blood without risking significant immediate and long-term injury to the brain is almost impossible.

Conversely, treating ischemic strokes can stop and even *reverse* brain injury. Cerebral ischemia is typically caused by a blockage in one of the cerebral blood vessels, most likely the result of a clot (a thrombus) in the blood vessel, which can form as a result of blood's natural tendency to clump platelets together with adherent proteins. The chances of clotting are higher when the flow of blood is slowed—the evolutionary mechanism to stop bleeding—or in hypercoagulable states where the platelet cells and adherent proteins are more prone to clump. Blood clots can also form in distant parts of the body, dislodge or fragment, and then traverse the circulatory system until they end up wedged in the narrowing of a blood vessel in the brain. This is an embolism, which is derived from the Greek *embolus,* literally "peg" or "stopper." Regardless of why blood flow is stopped, reestablishing blood flow before the brain cells begin to die from lack of oxygen is imperative in ischemic strokes. If blood flow can be restored soon enough, the affected brain cells may resume normal function, possibly reversing any neurologic deficits sustained during an ischemic stroke. Seconds often determine whether individuals return to their homes and their life's work or spend the rest of their lives in a nursing home—bedbound, unable to move or ask for help, and at risk of quietly drowning in their own saliva.

In Teri's case, doctors quickly performed a CT scan and, once they ruled out a hemorrhage, administered a dose of intravenous tissue plasminogen activator (IV tPA), a natural protein that helps break down blood clots, to reestablish blood flow to Teri's brain tissue. Administration within three hours of a patient's *last known well* is crucial, although a select subset of patients can safely respond to the protein for up to four-and-a-half hours. The best recoveries, however, most likely occur when a patient receives IV tPA within an hour of the onset of symptoms. Which is why Teri, now eighty-two minutes removed from her *last known well*, was shouting internally for the medical team to "*Hurry up! Hurry up! Hurry up!*"

As the tPA began to flow into her veins, a nurse told Teri they were transferring her to a "comprehensive stroke center" where she would most likely undergo an emergency thrombectomy, a mechanical clot retrieval, which required an interventional radiologist to snake a wire through an artery in Teri's groin all the way up to the cerebral arteries, corkscrew a hole through the center of the clot, open up a stent, then carefully drag the entire clot-stent amalgam back out of her body.

A few minutes later, paramedics loaded Teri into an ambulance. At least she was moving, she thought, in transit to a hospital she knew well. This at least was reassuring. But suddenly she discovered something new to worry about. A bump in the road caused her blood-tinged hospital blanket to slip off of her. Since handing her coffee to Parker, she hadn't paid much attention to her left arm, which was completely devoid of sensation. But when the blanket fell down, she discovered that the paramedics in their haste had accidentally pinched her arm in the stretcher. Growing there was a large hematoma, a collection of blood under the skin. The left side of her body had stolen off completely to the unfamiliar: numbed, bruised, trapped, helpless.

How can a stroke alter a person's physical and subjective realities in a manner that is so startlingly mysterious but so immediately discernible? The heterogeneity of strokes reveals the humbling complexity of the relationship between the brain and a person's behavior. Historically, uncovering the functional anatomy of various areas of the brain was made possible through the study of physical, stroke-related deficits. Nineteenth-century

French neurologists, many from the celebrated Pitié-Salpêtrière Hospital in Paris, pierced the opaque veil of the neurosciences through a combined expertise in anatomy, anthropology, and pathology. By studying the clinical phenomenology of patients and conducting careful post-mortem explorations of brain abnormalities, the medical field was able to gain a fundamental understanding of how the central nervous system operates, particularly which parts of the brain are responsible for which functions of the body. Jean-Martin Charcot described many of these clinical syndromes and their so-called "localizations" that represented the corresponding neuroanatomical site of injury. For example, *hemiparesis,* or weakness of one side of the body, corresponds to loss of functioning neurons in the motor cortex on the *opposite* side of the observed weakness. This localization becomes even more likely when there is *hemianesthesia,* loss of sensation on one half of the body and on the same side as the hemiparesis, implying damage to both motor and somatosensory cortices. The weakness and numbness to one side of the body may seem counterintuitive but are in fact due to how our nervous system develops. Neurons from the cerebral cortex, the outermost layer of the brain where the majority of neural circuit "processing" occurs, connect with other neurons through synapses to send signals through the basal ganglia, brainstem, spinal cord, peripheral nerves, and their respective final destinations in various tissues of the body to imbue them with life as, for example, motor nerves to muscle fibers or sensory nerves to the skin.

Because the left side of Teri's left face and arm were weak and numb, her *left*-sided hemiparesis and hemianesthesia were a manifestation of ischemia occurring in the *right* cerebral hemisphere. Her ability to see her entire visual field remained intact, indicating that there was unlikely any damage to the occipital lobes, the most posterior part of her brain, where her primary visual cortices are located. However, one need not be blind not to see. Teri may not have been aware that the entire left side of her body was missing. Not a blindness per se, but rather a total subjective nonexistence. Our capacity to notice and sustain regard for the left and right parts of our external world allows us to recognize, perceive, and process the stimuli of each part simultaneously. Attention dictates the perception of existence, the conception of the entirety of one's own reality. If we cannot attend to an object, that object—at least subjectively—does not exist. The profound deficit of losing the left half of one's reality is known as a *hemineglect* or *hemispatial neglect* and is unique to right cerebral cortical damage. The specific neuroanatomic areas involved include the right posterior parietal cortex or lower down at the boundary between the temporal and parietal lobes.

Our brain's ability to attend to both halves of the world relies only partly on the left-sided counterparts of these cortical structures for the *right* half of the world while the right-sided structures are relied on almost entirely to attend to *both* halves of the world. Injury to the right cerebral hemisphere can completely surrender the leftward universe into oblivion. Interestingly, these are the same areas of cortex with observed thinning in mirror-touch synesthetes like me. Perhaps this is central to the mirror-touch experience of blurred boundaries between the self and the other—not an alarming deficit found in left hemineglect from a stroke, but a sort of neurologic self-neglect. Or, in gentler terms, a tendency to forget the self, to reflexively replace the *self* with reflections of the *other* while failing to automatically process or attend to *both*.

These attentional networks are integrated so deeply with our other neural circuits that people who experience a hemineglect can lose awareness of the fact that the left side of their world is missing. They often fail to recognize that their deficit even exists (*anosognosia*), with or without the ability to experience an emotional reaction toward the neglect (*anosodiaphoria*). People in this condition will involuntarily confabulate in order to resolve logical conflicts. For instance, if you were to lift the paralyzed left arm of a person suffering from anosognosia toward their face, all the way over into their right visual field, and ask them, "Whose arm is this," they might look at you puzzled, wondering why you might ask such an obvious question, and reply, "Yours." Known as *somatoparaphrenia,* this delusion of lack of ownership over the left side of his or her body can lead some to believe his or her left limb belongs to a family member or an unseen stranger, even if neither person is present in the room. Their writing and drawings lead to the cramming of scribbles to the right side of the page. Clocks, for instance, are drawn with the numbers all scrunched to the right side, between 12 and 6. Sketches of flowers will appear as denuded daisies battered by bitter west winds.

Occasionally, the first sign of a hemineglect will be slightly more isolated and much more perplexing to friends and family, such as suddenly eating only from the right half of the plate or shaving only the right side of the face. John Hughlings Jackson, from the cadre of neurologists gathered at Queen Square, London, made the first comprehensive documentation of hemineglect in 1876, which he described as an "imperception." This imperception can include an inability to form whole mental representations in the mind's eye, as demonstrated by E. Bisiach and C. Luzzatti, who in 1978 asked patients with hemineglect to describe from memory the Piazza del Duomo from the viewpoint of the cathedral at the center of the square. These patients would often unknowingly leave out all streets and buildings located on the left side of the square. When asked to then visualize the square from the viewpoint looking straight-on at the cathedral, the hemineglect patients would then exclusively describe the streets and buildings on their right-hand side, which they had failed to mention the first time.

Because Teri was so fixated on the emergency room clock, she purposely turned her head and focused her eyes left. This inadvertently led the medical team to suspect that she either had a very large stroke or was experiencing a seizure. Epileptic activity isolated to one cerebral hemisphere can often cause a deviation of the eyes and head toward the *opposite* side of the actual electrical activity. Meanwhile, in stroke, there is an expected decrease in electrical activity from neuronal ischemia, thus the head turn and gaze would be toward the *same* side of the lesion due to weakness rather than overactivation of muscles from seizure.

The other neurologic deficit that localized Teri's brain injury to the right cerebral cortex was her abnormal speech. A speech abnormality from a stroke can appear in a patient as the inability to comprehend speech, though the patient can still fluently produce words that sound like a nonsense foreign language to others. This is called Wernicke's dysphasia or aphasia. It involves the part of the cerebral cortex called Wernicke's area, which is located in the posterior section of the superior temporal gyrus at the intersections of cortical areas responsible for auditory, memory, and visuospatial processing. Alternatively, the inability to produce words with a retained capacity for understanding other people is often the result of an injury to Broca's area, a Broca's dysphasia or aphasia. Paul Broca provided the first anatomical proof of this in 1861 at the University of Paris. Patients suffering from the inability to produce words, he found, had a lesion affecting the pars triangularis and pars opercularis of the *left* inferior frontal gyrus at the intersections of cortical areas tied to executive functioning and the motor activity of the mouth.

Broca gained his insights from studying a patient he nicknamed

"Tan" because *tan* was one of the few words he could say. Following an autopsy, Broca discovered a syphilitic lesion in Tan's left frontal gyrus. Tan was right-handed. Broca's research found similar patterns in other patients like Tan, which led him to conclude that a person's handedness tended to indicate a dominant language area on the *opposite* side of the brain. Although we now know that most language areas are left-sided regardless of handedness, there are some cases of right-sided language areas in left-handed people. Because Teri is *left*-handed, if her speech difficulty was truly a Broca's aphasia (suggested by her retained ability to understand the language of others and follow commands), then it could have been consistent with brain injury to her right cerebral hemisphere affecting her relatively rare *right*-dominant speech areas.

Ultimately, Teri's symptoms all pointed toward right hemispheric cortical injury involving posterior portions of the frontal lobe, extending along the parietal lobe, and likely involving the edge of the temporal lobe. The smallest common denominator between these areas is that they all receive their blood supply from the right middle cerebral artery. Therefore, along with the fact that she had an almost instantaneous onset of her symptoms, the highest likelihood diagnostic scenario was that she had a clot wedged in her right middle cerebral artery, and this clot cut off the blood supply to the parts of the brain supplied by this artery.

Meanwhile, in the back of the ambulance, Teri's hematoma continued to grow. Unable to speak, she could not ask the paramedics to free her left arm. Her nursing instincts reignited. In the absence of pain, she tugged at the left arm with her right hand until it was pried free. To prevent the hematoma from expanding, she applied direct pressure on it. Surely the size of the hematoma was a harbinger of the tPA's activity. The "clot-buster" was thinning her blood. She pushed down harder.

As Teri applied more pressure, she noticed a change. It was

faint but definite. Off in the distance of her senses, she could feel the rising crescendo. Sensation was returning over the area of the hematoma. A surge coursed through her body. Not like electricity, but as if blood flow was returning to the left half of her body, just as the tPA finally dissolved enough of the clot in her right middle cerebral artery. Teri took in a slow deep breath.

Her terror abated long enough for her to take inventory of her body from head to toe. She was methodical in her scan through the use of the most standard way of measuring the severity of an acute stroke, the National Institutes of Health (NIH) stroke scale. Her vision felt normal as she looked around the back of the ambulance. IV bags swayed and stowed toolboxes shuffled with a muffled jangle with each bump. She lifted her left arm and, though weak, she could raise it up in front of her as she watched it slowly drift in the air in front of her like a half-filled balloon. She was able to elevate her left leg for five seconds and, as she passed her right hand across her left arm, she realized it felt as if it were covered in a thick wetsuit. The sensation was still dim, but it was no longer the shadow of a disembodied left arm. To her relief, her left arm felt as if it had merely fallen asleep. Her vision suddenly blurred. This was not a new neurologic deficit from her stroke. Her eyes welled with tears. She knew she would be able to embrace her husband and her son again with both arms. Even if she could no longer speak, her love would not be shackled. She took in another deep breath, then exhaled.

Next, Teri needed to assess her speech. She opened her mouth and with just the slightest amount of effort she was able to give birth to language once again. "Ma . . . Ma. Tip . . . Top." These were words provided in the NIH stroke scale handbook for physicians and stroke nurses to assess language. The paramedic, not used to measuring the NIH stroke scale, looked back at her from the passenger side seat. "Fifty . . . Fifty. Huck . . . Huckabelly . . .

Huckle . . . Huckleberry." The paramedic raised an eyebrow—half surprised that she was speaking again, half piqued that perhaps this woman in the back of the ambulance was actually suffering from psychosis rather than a stroke. Teri attempted to speak, but the paramedic's expression did not change. She knew that her speech remained slurred and incomplete. The paramedic asked her a question, but she refused to answer again. Seeing that only every third or fourth word would be understood anyway, she chose not to speak. She did not answer with words again until she felt that she would be understood. There was security in reclaiming this control over herself. In her moment of extreme vulnerability, *this* she had authority over, and she knew it was worth holding on to.

The last reliable memory that Teri had that day was seeing the familiar face of a nurse she once worked with when she arrived at the hospital. Her memories move in and out of focus. She remembers a nurse she had never met before studying her face. The nurse turned to look at Teri's husband, then turned back to Teri's face. "Is her face always like that?" she asked.

Over the next two days while in the neuro-ICU, her blood pressure and heart rate traveled through the extremes of urgent ICU management. As a runner, her circulatory system was conditioned to run at extraordinary levels of efficiency. An average systolic blood pressure is somewhere between 110 and 120 mm Hg. An average heart at rest beats about 60 to 100 times a minute. Teri's systolic blood pressure resides in the low 90s. Teri's heart rate? The low 50s. In the setting of the acute stroke, her blood pressure had increased to about 130 to 140 mm Hg, several standard deviations from her typical blood pressure. In the setting of ischemia to the brain, the rest of our bodies evolved to respond by increasing blood pressure in order to increase cerebral perfusion pressures and keep as much of the brain alive as possible. In the first hours to days after stroke, the body will begin to readjust how it manages its blood pressures as it calibrates alongside the variable blood pressures within the cerebrovasculature, which are normally proportional but not exactly the same as the blood pressures of the rest of the vascular system. During the first night in the ICU, as Teri helplessly succumbed to exhaustion and entered sleep, the activity of her parasympathetic nervous system would dilate her blood vessels and slow her heart rate. Her blood pressure would dip into the high 70s, and her heart rate would gradually slide down to the low 40s. A systolic blood pressure in the 70s for most is cause for emergency. But because Teri showed no signs of any neurologic changes, she likely was not depriving her brain tissue from being perfused by blood. Yet, given the concern to prevent extension of the brain ischemia, the physician and nurse caring for her wanted to start her on hemodynamically active medications that can increase blood pressure and heart rate in a manner so sensitive that it must be slowly and constantly infused through an IV drip. Teri, able to speak more intelligibly, looked at the nurse and pleaded, "Please don't do that. You'll never get me off of it." It was a reasonable request if one is using the neurologic exam to guide management of her hemodynamics but unreasonable if going strictly by the numbers. Above x = abnormal. Below y = abnormal. Clinical management seems straightforward if treating reference ranges as rules but, if we are to look beyond overgeneralized parameters averaged across many, the specifics of a dynamic clinical context and a person's physiologic milieu must prudently take precedence. Teri's nurse played the messenger for Teri and her physician. They agreed that if she were to develop any worsening of her symptoms at all, even the slightest increase of slurring or a drooping of her face, they would start her on the blood pressure medication. The nurses came in the next morning to visit her and joked, "We were all just sitting there watching your monitor. Whenever your rate would get to like 42, we would go and just kind of touch the door so you would wake up a little. Your heart rate would go up and the alarms would stop going off."

On her second day of the ICU, Teri had an MRI performed for better resolution of what parts of her brain were damaged. As expected, she had clearly demarcated areas of ischemia within the distribution of the right middle cerebral artery, but fortunately there was no evidence of bleeding or ischemia encroaching on other areas of brain tissue. There were also no other "silent" strokes evident throughout the brain, which by definition do not manifest with identifiable deficits on neurologic testing, suggesting that whatever embolic event had occurred was likely associated with only one traveling clot that lodged itself into the right middle cerebral artery and had not showered the rest of the brain with additional strokes.

Teri had a transthoracic echocardiogram performed as well, which involves the use of ultrasound to create a picture of the heart and surrounding vessels to look for intracardiac thrombi or another source of the clot in order to help guide the next course of action, which is first to prevent another stroke from occurring and second to recover as close to her usual state of functioning as quickly as possible. Teri's echocardiogram, however, did reveal an abnormality. There was a patent foramen ovale (a PFO). Colloquially referred to as a "hole in the heart," though this phrase can mean many other variants in cardiac structure that may not be pathological, a PFO is a small malformed opening between the walls of the heart that allows blood returning from the body to seep through the walls of the heart and bypass the lungs. Thus, if Teri had a blood clot in one of her calves that fragmented into smaller emboli, which traveled all the way up into her heart, one or more of those emboli could have skipped the lungs and entered into the left atrium of the heart through the PFO. The heart could have then pumped a clot or clots into the rest of the body with the likeliest of places being the brain, which is a straight shot above the heart through the arteries of the neck.

A physician returned to see Teri after the transthoracic echocardiogram to update her. Based on the results of the study, he explained that she had a PFO, telling her, "We're just going to put you on aspirin. Go run your marathon. You're fine." Teri was already taking aspirin. Her primary care physician started her on 81 mg of aspirin, a baby aspirin, as soon as she turned forty years old. Teri did not speak up. Instead, she shut down. Teri remembers feeling hurt at first, but she thought, "You know what? This is fine with me. I don't need to advocate for myself here. I live in the world of clinics and hospitals and I know doctors who will take care of me the way I feel I deserve to be taken care of. Thank you, and goodbye."

Teri stroked on Monday. Her first session of physical and occupational therapy was on Friday. At the end of the first session, she felt ragged. She sat in an uncomfortable upright leather chair with her left arm in a sling connected to a pulley that she would hoist with her right arm. She wondered if the strength in her arm would ever return. She had previously worked with this physical therapist following a knee injury, though this was the first time she had ever seen her physical therapist feel sorry for her. Teri felt sorry for herself, too, at that moment. Just for a minute. She allowed herself that, then fought through it.

"Go run your marathon," she remembered the physician telling her. At that moment, she hardened her resolve. She would run her first marathon as scheduled. She would run her first marathon even if it meant holding up her left arm with her right while she ran.

Despite her determination, Teri had to battle exhaustion, periods of low physical activity, a general feeling of slowness and a frustrating sense of weakness as she prepared for the marathon—not to mention weight loss, due largely in part to the fact that her stroke had made it more difficult for her to swallow her food. Weight loss is a cardinal feature of frailty, a state of increased vulnerability in reserves across multiple physiologic systems. Frailty after stroke impairs a person's ability to cope with daily stresses, expected or unexpected, and decreases chances of survival while increasing the likelihood of developing stroke-related complications. Though typically neglected, frailty is also closely tied with a person's psychosocial circumstances, such as social support from one's tribe.

The mechanisms through which psychosocial factors can affect recovery are often as complex as the neuroendocrine and inflammatory systems at a molecular level or as straightforward as having friends or family members available to coach and practice physical and occupational therapy at home. Over dinner, for example, Teri's husband, or Parker, might casually make the suggestion "Let's eat with our left hands tonight." They were mindful of Teri engaging her left arm. If they didn't remind her, she told me, she would absently tuck it in her pocket or on her lap, weary of banging it or getting caught in a closing door. Often, she would forget that she even had a left arm. The arm felt too much like oversized carry-on luggage—heavy and cumbersome, something *other.*

As trite as it sounds, "use it or lose it" is an apt description of what can occur if paretic limbs are ignored. As they become less supple, tendons and joints form scar tissue called contractures. Without signals from the brain to activate muscles, muscles will initially lay limp, but, over the course of few weeks or a few months, the disinhibited signals from nerves originating in the spinal cord lead the muscles to insidiously contract, cramp, spasm. The continuous decrease in muscle tone eventually leads to spasticity, and spasticity leads to even more contractures. Without use, the appendage slowly draws closer to the body as a rigid fixture.

Teri experienced a taste of this on a twelve-mile practice run.

Only a couple of weeks removed from her stroke, she was nearing the end of her eleventh mile when her left arm and fingers started spasming, pulling closer inward. Her arm became almost as numb as it had been on the day of the stroke. She pushed to the end of her practice run, plopped down on her couch, took a few slow deep breaths, and hydrated. Two liters of electrolyte-enriched water later, sensation returned to her arm and spasticity slowly released its grip on her muscles. Still alarmed though, she called a close friend, a neurologist and a medical director at the stroke program Teri coordinated. They talked about her neurologic exam, confirming that Teri hadn't experienced a setback and that her new baseline functioning remained the same, which gave Teri a sense of relief.

But, at the same time, Teri understood that her desire to run the marathon confounded judgment. Like her neurologist friend, she knew that dehydration and stress on her cardiovascular system might cause a drop in blood flow to her brain, which would lead to new ischemia, extending the area of her stroke. She felt selfish. But running the marathon had become a driving force in her recovery and her return to Teri, to reclaiming her body as her own. The motivation to recover has been found to be a reliable predictor of which patients will have greater motor and speech recovery when most other factors are held equal. Teri wanted to take advantage as much as she could of the neuroplasticity still active in her brain. She began to pay close attention to her body, the fluctuation of any symptoms. She focused on the hopeful prospect that she was building new pathways in her neural networks.

At the behest of her neurologist friend, Teri did see a cardiologist before the marathon. The cardiologist agreed with the decision for her to continue taking aspirin but also recommended she see a specialist to evaluate her PFO more closely. She interpreted the interaction as an implicit approval of her participating in the marathon. She went home and told the news to her husband. He expressed that as long as the cardiologist felt she could, then it was okay with him. She invited her husband to come join her at the marathon, but she already knew his answer, "No . . . Before the stroke, you had planned on a girls' weekend for the marathon. Let's just keep it the way you wanted." Teri appreciated his concession but also knew that he likely felt more assured in his decision by the fact that her best friend, who also happened to be her primary care physician, would be her running partner during the marathon.

Deep down, though, Teri still felt as if she were a "ticking time bomb." This sense of anxiety is common among stroke survivors who experience a stroke without a specific cause, so-called "cryptogenic" strokes, which actually make up a large portion of strokes. If the underlying cause is unknown, then it cannot be completely treated or eliminated, which leaves the chances of having another stroke about the same. Anxiety sets in. Patients will enter into the hyperadrenergic fight-or-flight state but have no way of managing it. Without finding a reprieve, people can't productively expend the stress response, which leaves them in a suspended level of stress with all of the attendant consequences of increased risk of cardiovascular disease.

I recall one diabetic patient who developed slight double vision that resolved but who was referred by his primary care physician to the emergency room for a neurological evaluation. The imaging of his cerebrovasculature revealed a tight stenosis of his basilar artery—the slender artery that runs up along the belly of the brainstem and provides almost all of its oxygen. His basilar artery was almost completely blocked. If that blood flow reduced enough or—worse—stopped completely, an infarct in parts of his brainstem would develop, rendering him almost completely paralyzed, taken hostage within his own body. These patients only maintain the ability to make slight (mostly vertical) movements of the eyes but have no true cognitive impairment, hence the moniker, "locked-in syndrome." To reduce the likelihood of occlusion from a blood clot or a worsening of the atherosclerotic plaque, he started indefinite anticoagulation and cholesterol-lowering medication. I followed up with him several months later. He shared the news that he was able to control his diabetes with regular exercise and major changes to diet. His hemoglobin A1c levels dropped from over 10 percent to 6 percent, the personal health equivalent of winning an Olympic gold medal. His wife outwardly celebrated his progress, but I felt oddly uneasy. I sensed an awkward restraint in my own body, much like the suppression of a smile or the smothering of joy. I paused and took a chance. "Tell me a little bit about how you've been feeling lately." There was silence for a few seconds. I felt swelling around my eyes. He began to flush. He came undone. All of the emotion, all of the anxiety, that he had been packing far down at the bottom of his psyche pierced the surface and burst forth. He wept and choked with heavy sobs. His forehead furrowed, his mouth arched into a deep grimace as he said with a wavering voice, "I'm a dead man walking." He was able to reduce his risk factors for a second stroke, but at what cost? He was driven purely by fear of his death, which he felt looming behind him, his chthonic shadow. If he had psychosocial supports to bolster his resilience, could this amount of physical, psychological, and existential suffering have been avoided? We moved on from talking about medications and began to talk about his life.

The day of the marathon, Teri's heart pounded in violent protest as she walked up to the starting area. The longest distance she had ever run, even before the stroke, was eighteen miles. She tried not to think of all the ways she was going to fail. She reminded herself that all she needed to do was focus on the race as individual splits at 5K, 10K, half marathon, twenty miles, and finish. She did not have to finish. She had already won.

Running alongside her running-partner/best-friend/doctor, she was surprised to find how absolutely amazing she felt while running, even at the sixteen-mile mark. She had just begun to feel that perhaps she would be able to finish the entire race when she felt a sudden change. Teri started to develop sharp cramps and spasticity, though not in her left arm. She felt the pain radiating along her entire right arm all the way up along her shoulder blade and neck. She thought to herself, "Oh my Jesus, I just had another stroke on the other side . . ." She effortlessly threw out a train of randomly chosen curse words interspersed with repetitions of "Oh God, I just threw another clot into my brain." Teri knew that her chance of having another stroke was already higher simply by having had a stroke at all. Though, if a first stroke to the right side of the brain is devastating, a second stroke to the other hemisphere—striking down the other half of her body—would be catastrophic.

Teri took a deep breath through her nostrils. She paused and remembered everything she had learned as a runner. "When a part of your body starts to hurt, whether it is your head, or knee, or toe, start to take inventory of your body from top down." Much like a meditative body scan, Teri began assessing her body's alignment and worked her way from the top of her head down. It was not long before she realized that she had tucked her left arm inside her fuel belt, a belt where quick, convenient, easy-to-swallow foods were stored. Because her left arm was resting quietly, tucked and completely immobilized out of sight, her poor right arm was pumping in double-time, swinging twice as hard. Teri's friend noticed that Teri's pace began to slow. "Teri, are you okay?" Teri pulled her left arm out of the fuel belt, "I am now."

Teri then reached into her fuel belt and pulled out a ball of putty. Three hours into the run, Teri reassured her friend, "It's time to do some physical therapy." She began to do exercises with the left and the right arm, alternating her grip with the putty, changing the position of her wrist. She pulled out the extra packets of high carbohydrate "goo" she collected at the first stop to create more variety for her left arm exercises. She even began to do exercises with her face, winking back and forth, smiling with variations, and pouting her lips to make duck faces. Within a mile, her body had settled into an even pace again. The pain resolved.

At mile eighteen, Teri jogged closer and closer to the medic tents. This was her chance to forfeit the race. Instead, she began to cry. She realized that she wouldn't have to forfeit the race. She could act like a marathon runner and no one had to know that she had just had a stroke. Her lower lip quivered in a profound surrender. Teri finally allowed herself to feel *normal* again.

Three miles from the finish line, Teri had yet to feel the proverbial "wall" that marathon runners often describe between miles seventeen and twenty. Knowing that she had the equivalent of a 5K remaining with a personal 5K average time of twenty-four minutes, Teri started to increase her pace. She thought to herself, "Oh, it's a 5K, it's not a big deal." As she continued to increase her pace, however, she suddenly couldn't breathe. She was struck with fatigue. The phrase "I'm going to have to walk" invaded her mind and looped over and over. Teri took a deep breath once again. She returned to her strategy of taking a mindful inventory of her body. She looked down at her watch and realized that she had actually begun to run seven-minute-and-fifty-second miles. Teri's friend looked at her, "I wasn't going to say anything, but I was wondering if we were going to just go ahead and sprint to the finish!" They slowed back down to a more sustainable nine-minute-and-thirty-second pace. Once again, Teri was able to settle back into her own body.

At mile twenty-five, Teri became quiet. She felt as though there were no more fuel to burn in her body. She had used all the energy gel in her fuel belt. She pushed on. Her steps were now being powered by throwing her very body, piece by piece, into the fuming furnace of tenacity driving her toward the finish line. The voice in her head had also grown silent. Perhaps the blood flow toward Teri's frontal lobes had decreased, leading her to experience the hypofrontality of the "runner's high," where all is novel, all is meaningful. One's attention is overcome by a disinhibited mindfulness where the ability to perceive and attend to all the details and sensations of the external and internal world becomes gloriously expansive. This is where the seemingly demonstrative demarcation between spirituality and the physical body fade into one another into a subtle, bottomless haze of consciousness. The sound of people cheering along the sidelines reverberated throughout Teri's body. The trees lining Lake Superior took on a vibrant shade of emerald she had never seen before. Time slowed. Teri did not need to walk.

Sure, the left side of her face drooped the way that it had on the day of her stroke, and she had to push her words out with force once again, but on Day Twenty-Six she ran twenty-six miles, her first marathon.

Yet the unpolished stories of our lives usually do not end with a final glorious triumph, a blazing victory followed by a roll of credits. Our stories go on. More likely, more marathons remain beyond the finish line. One more thing, one more time.

Since the stroke, Teri has allowed herself to keep taking one step at a time. A space emerged between every action and every thought, a sudden pause that opened up her life after years spent trying to control every moment. She allows herself to take as many deep breaths as she needs to take inventory of her body and her life. No longer feeling the need to push ninety-hour work weeks, she allows herself to be at ease as long as she is true to what her mind and her body need.

Whenever I spoke with Teri, I frequently felt the sensation of light brushing on the right side of my lower face that mirrored Teri scratching her face. Teri had developed the habit, mostly to hide the asymmetry of her face. Growing up, she had taken so much pride in her smile. It was her signature. Having an asymmetric smile left her feeling permanently marred, crooked. To regain her old smile, she practiced smiling exercises in front of a mirror. Twenty minutes every night, she would face down her new face, the new person looking back at her. For the first few weeks, as she stared back at herself, Teri felt her skin crawl. Little by little, though, she got reacquainted with herself, with this new person in the mirror. She grew to appreciate her company and eventually surrendered to her, too. She embraced her lovingly as her new normal.

Through her stroke, Teri came into her own. She quietly rewrites the story of who she was into the story of who she is now. The inspiration for her verses came in moments so small that had she not been attentive they might have fluttered past her. For instance, Teri played golf for several years and believed she was absolutely helpless at the sport—constantly attempting to push and control the club, unable to achieve loft or height to her drive. She told herself she was limited in her ability to estimate the geometry of height, depth, and angles. But after her stroke, she decided she was not done with golf and took lessons from a dear friend she could trust, a member of her tribe. She was amazed at the way her brain, previously dominant to one side, had now adapted to a balance between the use of both sides of her body. She can appreciate the subtle movements and sensations of both arms. She is able to connect with the angles of her drive and, by lightening her control, she can follow through her swings and drive further than ever before. She can allow this newfound gentleness to strengthen her.

The inspiration for Teri's new verses also enters her life large with unmistakable weight and grandeur, trumpeting a new chapter. Teri had no idea that running a marathon twenty-six days after her stroke would make her world as big as it became. Only one hundred days out from her stroke, she found herself getting ready to give a talk to over three hundred people in Dallas, Texas. The morning of her talk, she wandered on the grassy knoll of Dealey Plaza. The sunlight cast her shadow in front of her. She noticed that her shadow did not have a facial droop. Her shadow did not have a paretic arm. Teri recognized that this was a "moving forward day."

As Teri shared her story at conferences and online, she emphasized the importance of making sure that everyone can detect a stroke. In her red dresses, she has made it clear that it is unacceptable that stroke is the third leading cause of death in women but only about half of them know at least two signs of stroke (facial weakness and abnormal speech) and only a quarter know there can be other more unusual symptoms, such as dizziness, headache, or numbness. Teri tells her story though brief social media posts, sharing the signs of a stroke on her Facebook and Twitter accounts daily. One night, she was awakened by a text message. It was from a high school friend she hadn't spoken to in twenty-five years. The text message simply read, "Because of all of your annoying posts about the signs and symptoms of stroke and getting treatment, I got my twin sister to the hospital and she got tPA in time. Because of your story, I saved my sister's life tonight."

While we might be naturally driven by the desire to win,

each step we take regardless of the outcome is but one more step to propel us in our story. Whether forward or backward, no step is wasted. From neuron to momentum, from rise to fall, Teri appreciates the experience of each step, midstride, as an act of grace. The next step will be an even greater miracle, another marathon—as we continue to keep stride.

CHAPTER 9

The Visions of Anekantavada

STEPPING INTO MY FINAL YEAR OF FORMAL CLINICAL TRAINING, my behavioral neurology and neuropsychiatry fellowship, I was struck by a palpable change in my interaction with patients. Compared to the previous four years of my residency, I had more time—much more time. I relished in its wide open, almost pastoral, expanse. The additional time allowed me the opportunity, the freedom, to interact with my patients in full, to make sense of my mirror-touch sensations. Fiona's surrendering, Rosie's storied tribe, Teri's new normal—they all helped me in this endeavor, allowing me to apprehend with all my senses my own perception of the world and gaining a full-bodied understanding and emotional appreciation of other people's perceptions and points of view.

As a result, I was able to more fully focus on my patients' individual stories, their case histories and private experiences. Listening in a relaxed and open setting, gently attending to us both through careful observation and mirrored sensations, allowed me to see things with calm and attention, as if peering through the lens of a naturalist. Over time, I started every patient visit with a semistructured interview, making it clear to him or her where we were going. Then, after building a strong rapport, I posed more personal questions, trying to dive deeper into the patient's story because, as I learned, the crucial topics—the most pressing issues—often lie behind years of erected barriers, cemented in place by silence, avoidance, and embarrassment. Together, we came into contact with the places that hurt the most, one of the most tangible forms of insight available. Asking a patient—or as I also learned in my personal life—where it hurts, reveals unspoken but pivotal parts of the story, the troublesome parts we are often unable or, for several cognitive or emotional reasons, unwilling to admit in fear that these stories will come to define us, limiting our ability to escape or transcend them in the process.

It is no great revelation to say that circumstances and conditions are too often beyond our control. But as narrators of our own stories, we almost always possess the power to control how—and with whom and when—we talk about the circumstances and conditions out of our control. During their telling, these stories become ours and ours alone. In other words, by owning our own narratives, we choose how to live through them. And living through them rewards us with new perspectives. My younger brother, Rainier, for example, was born color-blind. While the retina's light-receiving cells usually react to three wavelengths of light—short (blue), medium (green), and long (yellow-green or red)—Rainier's retina only reacts to two: short and long. His red traffic lights are probably closer to what others call orange, while his greens are someone else's whitish-gray. If a bunch of green Granny Smith and red Pacific Rose apples were mixed in a basket, they would appear to him as different shades of grayish-yellow; he would be unable to tell the difference between the two until he took his first bite. And yet, he still loves color. He wears loud bursts of colors that he can easily discriminate on his body, mixing and matching items of clothing according to the colors he sees rather than the colors other people claim to see. Because they help him match his perception of colors with colors perceived by other people, blue and yellow remain his favorite colors. Rainier embraces his color blindness as a gift that reminds him, time and time again, that he comes with a distinct experience, a unique perspective. But there were, of course, times he felt excluded from shared experiences when confronted with this difference, this distance, between him and other people. In high school he went on a field trip to an art museum with his class. The tour group stopped in front of an oil painting on the wall. His classmates exclaimed at the beauty and intricacy of its colored designs although the shades they loved the most were imperceptible to Rainier. He instead saw images that were altogether different from what others experienced. He asked the rest of his group why they saw what they did, but none could articulate what Rainier was missing. None could imagine a reality where red and green could not be told apart from one another.

Over the course of my clinical training, especially my fellowship, I often returned to the thought of Rainier in front of the painting, how far away he must have felt, how painfully excluded. This was a familiar feeling for me and many people I've met, this constant feeling of *almost*—almost finding a common understanding, almost joining with other people in a shared experience, almost but never quite able to cross the final threshold of connection. To help my brother make this connection, I went in search of a pair of glasses. These glasses, designed to create optic filters through computer modeling and perceptual psychophysics, simulated in a color-blind observer the experience of perceiving colors he or she had never witnessed before. Perhaps, I hoped, I could help bring Rainier a little closer to a shared experience with these glasses. I wanted to give him an opportunity to experiment a little bit and explore a world filled with distinct hues of red and green—a world more aligned with everyone else, including me, his brother. I saved up some funds and presented a pair to Rainier as a gift.

Maybe it was distance, or perhaps we just had a hard time figuring out what the other might want, but Rainier and I hadn't exchanged gifts in many years, probably since I left for college. Rainier was surprised, both by my gift and what it promised. He didn't know what to expect. Instead of putting on the glasses right away, he decided to wait to wear them at a more interesting, a more colorful, environment. I asked him what he had in mind.

"A rodeo," he replied.

My father had grown up accustomed to seeing impromptu rodeos in Nicaragua, and these rodeos—cowboys and all—reminded him of where he was born and how America had gradually become his more native home. Because my father and mother were already going to the rodeo, they invited me and Rainier to join them.

Sitting in the crowded bleachers, as rodeo clowns danced around four-legged Plummer and Charbray bulls in the ring below us, Rainier carefully removed the glasses from their packaging. I could feel the thrill of his excitement on my own face: outer corner of the eyes crinkled, shoulders raised and tensed, jaw open lightly, brow furrowed slightly, both of us feeling as though we were about to cross a threshold into a parallel dimension. Would it tear open a hole into a new spectrum of creativity for him? Would nature still feel natural to him?

Rainier carefully slid on the glasses. I felt the hard plastic edges rise up along his nose, the snug tuck behind his ears, against his cheeks. As his eyes slipped behind the gray-tinted lenses, I felt the tightening around his eyes as he adjusted his vision. I felt his face go still, a quiet, hopeful countenance.

From behind the glasses, tears began to stream down his face.

"What do you see?" I asked.

"I . . . I don't know how to . . . how to *understand* right now."

The rodeo pen, which had been dim and difficult to perceive, had suddenly transformed into a vibrant red railing.

"Everything looks different."

He could not comprehend how bright the world was. He saw so many different shades of reds before him, bucking in his brain. He tried to describe them all to me, but he could only call them "red," though he acknowledged that every shade was different, richer in parts. He couldn't do the colors justice. Nor could he describe in full the new vibrant colors he spotted on the clothes of the people around him. Blues were suddenly more vivid, and everyone in the stands seemed to have donned, in the moment he slipped on the glasses, redder versions of the same clothes. Trees. The trees were a different shade altogether. He was overwhelmed and remained speechless for several minutes at a time, examining the world as if he were just delivered into it.

"The blue shirt . . . on this guy over here," he pointed to someone a few rows nearby, "It's so bright . . . It's like I'm watching a movie . . . Everything looks different . . . It's so hard to explain."

He pointed at a woman in a hoodie a few feet away, "And what is that color that she's wearing?"

"That's violet," I told him. "It's a mix of red and blue."

"So, that's what violet is . . . It's so different than I thought it would be."

I felt my own emotions welling up alongside my brother's, reflecting, feeling inside my chest the sensation of blossoming. A surge of new experiences arose from the throat and emerged beyond the eyes. It was as if each eye were a Primavera bud beginning to bloom, taking in all the light around it, feeding and enriching, free from all constraints, growing in great leaps.

SEVERAL WEEKS LATER I ASKED RAINIER WHETHER HE FELT A difference in his day-to-day life. "You know, it's helped me really take a step back and question why I have a certain action or reaction based off of a thought. It helps me hold back and take things in without judging. And, without a premature judgment, being open and accepting. Even in times where I may have an opinion or I'm in the middle of talking, I can take a breath and I can try to understand before I try to act or speak."

The Sanskrit term *Anekantavada* refers to the multiplicity of viewpoints where truth and reality are perceived differently from diverse points of view and no single point of view is the complete truth. I feel the echo of this word when I reflect on Rainier at the rodeo. Seeing the world with a new perspective is possible. It can be done by living an embodied life through our perception, channeled through each of our individual set of senses, and allowing our brain to receive them. By becoming more alive to our senses, we can expand the boundaries of our own consciousness in order to become more aware of the world around us. Through our senses we can appreciate that our mind and body do not simply work in tandem; they are one and the same. Our brain is a part of the physical vessel we call *body.* Like pigment is to color, the body is to the mind. Pigment is physical; color is mental. For me, the paradoxical space between these two reminds me of what I experienced with David and Ramachandran when they drew a 2 on top of a W. Their point of intersection felt like a contradictory expanse of opposing synesthetic colors, simultaneously revealing the limitations and the potential of the human brain.

Through Rainier's experience, I appreciated that my own

perspective, my own reality, was not only unique and unto my own but also utterly fragile. In a single unpredictable moment, my entire reality can immediately become unrecognizable, someone else's. Even when I am confident that I have a firm handle on my own perspective, through my experiences with Jordan I found that a new perspective, a world I had not expected to see first-hand, is forever only a moment away. Such a change isn't always pleasant but it is, more often than not, necessary.

As the winter of my fellowship year approached, amidst all that I had been learning from my patients, I neared the completion of years of structured clinical training. I would soon have more time to be at home with Jordan, so we could continue to learn from each other. I entered into the relationship vowing that I would avoid repeating what happened with Cristina, that I would not allow myself to get entangled too deeply in the other. I avoided the idea, even the language, of ownership of one over the other or of us merging. With our wedding vows, for instance, I insisted that rather than performing the symbolic ritual of lighting a single unity candle we instead plant two sequoia seeds to represent how we would grow together—while remaining separate.

Over time, however, I became more comfortable and less attentive to the possibility of losing myself again. One day I awoke and realized that Jordan was no longer just this other person I lived with and was legally related to. Jordan had become a part of my body, a phantom appendage to my mental body map. When he was away from home, traveling for work, it was as if an arm had gone missing. When he came back, the map became whole again. I hoped that he would also be this other body where I could find physical affection, to physically embrace me in return—not always, not unconditionally, but just enough. This was what I thought.

Around this time, he and I were considering having a child.

While trying to arrange our finances, deep in an expectant haze of impending fatherhood, I discovered in a single moment that Jordan had kept secret years of deception. Learning this forced me to acknowledge, once and for all, other truths about Jordan and our relationship that I had never let myself acknowledge—the pernicious humiliation of emotional and verbal abuse. To protect myself, consciously or subconsciously, I had sought refuge in work, in my patients, other people's stories. But I could no longer deny the truth. There I stood in this new reality, my eyes stinging, my ears ringing, wondering how I had let things go on for this long. My trust in Jordan was irredeemably violated, even if this trust were just one more illusion of perception, someone else's reality. I had to take back the control of my own story. Out of compassion, truly for each other, I made the decision to pursue a divorce, a word that I—like I imagine anyone else—never thought would ever be associated with my life.

As painful as it was, the divorce was essential in helping me once again disentangle. The entire process was trying—physically, financially, socially, mentally, and emotionally. It required me to seek out ways to literally find my own reflection to ground myself, to prevent from getting lost again entirely. This was a fall. A moment of intense grief and suffering. I took the time to sit in that grief, with the onyx sheet crushed into a tight mass lodged in my chest, to develop an understanding for why my body gave rise to the grief in the first place and why a word or an object might trigger a cascade of distressing thoughts and emotions. Rather than shelter myself away from grieving with unhelpful rationalizations or self-serving intellectualizations, I created space to appreciate and explore such questions without depending on them to rescue me from the pain of human suffering. My stomach ached, my sleep was disrupted. But, even though my pain was born through my body, through my brain, these feelings were not my fault. They were an essential part of my programming, a part of who I was and a part of who I was becoming.

Our built-in emotional systems can hijack our executive control networks so aggressively that we might not always be aware that we are relieving an unrecognized itch with an excoriating scratch. This is why building resilience is vital. It helps to create reflexive mechanisms to experience, notice, and reflect so that we may be able to think and feel critically and find all the ways to overcome the more challenging vicissitudes of living. Fiona, Rosie, Teri, Rainier, and many others helped me learn these fundamentals. I had thought of myself as a mildly resilient person but, echoing their experiences, I was surprised to find below the surface a quiet store of resilience that had been enhanced and bolstered to a new unexpected level, well beyond what I could formerly recognize.

Within the first few days after deciding to divorce, just the word *divorce* entering my mind made my stomach turn and recoil. I wanted to avoid that nauseating revulsion and stow it somewhere I wouldn't have to see, hear, or feel it, though my reaction to *divorce* was minor compared to my reaction to the word *abuse*. Neither my mind nor my mouth could get beyond the first syllable. I treated the *b* sound in *abuse* as if it were a land mine harboring the emotional explosive force of a bomb blast. The word set my entire limbic system on fire. If I pushed beyond the *b*, immediately the area around my eyes swelled, my eyes watered, my nose began to run, the hair on my skin would attempt to hide away in me, the onyx sheet—now a hardened ball of shrapnel—twisted from side to side, burrowing deeper into my center. In those first days, I thought of Fiona. My instinct was to fight against all the emotional chaos blaring from my limbic system, to gain control over it. But the more I tried to bend it into submission, the louder the emotional noise grew and the more I recoiled from the pain.

It was when I thought of Fiona, though, that I chose to gather every bit of courage I had in order to let go, to have faith in my body. The emotional noise was how my body, my brain, needed to scream and expel the distress and pain as if cauterizing a bloody amputation.

I let the words—divorce, abuse—come in without forcefully pushing them away. I listened intently to these four syllables, these twelve sad letters and all the pain that came with them because they were telling me exactly what I needed to know: where it hurt. And festering behind that pain was shame, guilt, anger, frustration, and, above all, fear. And yet I faced them head-on like a Kayapo in mourning. Taking my insides and pushing them outwards, I made space for calm. In those first days, I allowed this to occur until the day when the word divorce appeared and the revulsion was less severe. And, eventually, even the word abuse could enter in. I still felt my limbic system warm, but its heat came on more slowly, rarely overheating. This was far more manageable.

To facilitate this, I reduced my usual workload after clinic hours when I typically worked on papers and other projects. But, after some time, I felt that perhaps I could return to my previous level of productivity. I considered that it might help take my mind off the divorce to dive back into my work. I remember working late into the twilight hours that same night. I had a full day of clinic the following morning, a clinic that usually spilled over into the late hours of the early evening. I slept about four hours but felt somewhat refreshed. I drank a large coffee to make sure that I'd be as awake as possible throughout the day. In each patient encounter, I listened as intently as I could. But I found that as the patient and family's emotions would heighten, the mirror-touch sensations were less suppressible, almost out of control. My emotions heightened with theirs. I was overcaffeinated and sleep deprived, which only increased the intensity of these invisible sensations. I took what I had learned while in hospital emergencies and focused on one part of the patient's body, the collar or shoulder, or at the corner of the room, or drawing my attention into my own body into my toes, anything to keep the emotions and tears from spilling over. I was able to keep this up throughout most of the day, but then came the last patient visit of the clinic. We were running behind schedule, and the fatigue from the previous night had begun to get the best of me. I was moving sluggishly. Blinking and squinting, I felt a burning in my eyes. The last patient was a woman with dementia who had progressively worsened over the last year. Her husband was in the room with her. He was a man accustomed to keeping a busy work life. He had been spending more time away from home with work, and his frustration with the progression of his wife's dementia was clear. He was seething. His wife could now barely speak. I couldn't fathom the extent of his suffering, managing his home and all the pain that comes with having a spouse affected by a slowly progressing terminal illness. The woman had developed disinhibited exploratory behavior. She was grabbing things on the desk—the stapler, pens, paper, anything in sight—bouts of possessive impulses interrupted only by the enthrallment of her own shirt, or the strands of her hair, which she fiddled with obsessively.

When I finished the preliminary interview and performed my neurologic exam, I stepped out of the room to speak with one of the supervising physicians to present her case. As soon as I stepped out of the door, however, the husband followed right behind me, almost on my heels. I turned and saw nothing but aggression: his shoulders raised, brow furrowed deep, mouth grimaced, nostrils flared, fists clenched, chest expanding and contracting rapidly with shallow breaths, sweat beaded at his temples. He leaned toward me, over me, similar in height to Jordan. My heart quickened. I felt light-headed. Without warning he began yelling at me in crescendo, pointing his finger near my face, "That what you did right there, don't you *ever* do that again! Ever!" His voice shook with his rage. I felt the anger in myself, against myself. I felt the knotted bright scarlet threads tearing crimson and dark maroon coils through my arms and legs. Just as suddenly, the man stormed back into the clinic room. I had no idea what he was referring to. What had I done? What had I said that enraged him? I didn't think I said anything that I wouldn't have normally said with a patient. Everything in me was muddled, my own feelings and my own body. I felt as though I had just been taken back into the marriage I escaped.

When I got home, I had to let go and let the emotions pour forth. I was resistant to admit that I wasn't ready. It was only until I thought of what it would be like if I were talking to the younger version of myself, back in kindergarten in the playground, under the slide, that I was able to offer myself a kind hand, to befriend myself. This gave me the courage to choose what I needed to do for my future, which meant confronting that I wasn't able to work at such a rapid clip yet. I made the scary decision that night to admit that I was not ready, that these were not normal times, that if I really wanted to care for myself, as Fiona taught me, I needed to let go and surrender myself to myself.

I flew home to my parents, to that physical place that embodies an unshakable feeling of security where, during the most difficult period of the divorce, I could have support from my family. My parents helped with tangible acts like cooking meals and making sure that my physical needs were met. They were the people whom I knew with confidence that I could embrace affectionately whenever I wanted and always receive in return an equally loving embrace. As with many of us, after having left for college, the physical distance away from home had gradually become an emotional distance as well. I only spoke to my parents over the phone occasionally, deliberately prioritizing work and Jordan over them. When I came home at the airport, though, tears came readily when I saw my mom and dad standing at the passenger arrival area. They never failed to park their car at the airport and wait for me in person just outside the terminal. The feeling I had was similar to when I returned home for the first time after college for Thanksgiving, when I returned home from my trip to the Amazon, when I returned home to start medical school, when I returned home after ending my relationship with Cristina, when I returned home after my surgery with a shaved head and a beige sock over my head, when I returned home after the car accident covered in cuts and scabs with an arm sling and bloodied eyes. Here I was again, back home, back to my core tribe. A tribe that no matter how far away I was, gravity always guided me back to it whenever I needed it most. I hugged my mother, and again the limbic system heated, and I wept. I hugged my father, and I remembered the feeling of hugging him when he would come home from work, still warm from the Miami heat, his brown UPS uniform covered in sweat.

The next morning, my mom made breakfast. Over *café con leche,* I told my parents everything that had happened. In their presence I felt less of the scalding heat of my limbic system. Without recoiling, I was able to use all of my spoken language, English and Spanish. My mother and father looked at me with compassionate eyes that said they understood my pain, that they felt it. Through this pain, we were able to see each other, eye to eye, in a way we hadn't before.

Later that night, my brother drove me back to my parents' house from Miami Beach, the city where I was born. We sat in silence as we crossed over the causeway connecting the island of Miami Beach to the mainland. I looked out the window at the reflection of the skyline in the water. The passing streetlights along the causeway lit the inside of the car in a measured rhythm. Rainier broke the silence. "You know, if you want to talk about what happened, I'm here." He was the one who was now reaching out to me. At a loss for words I replied, "Yeah, it hasn't been easy." And then, bringing me in with his words, he told me quite simply, "Yeah, it sucks. I've been there before." It was a moment of connection. It wasn't about how exact his experience was to mine, how each detail of his past experiences mirrored the major and minor points of my own. This connection came from the fact that he had felt pain, too. At this simple ordinary moment, I felt as though we were suddenly tied closer—not just by blood, but by pain, which is often thicker than blood.

THE MORNING JORDAN AND I APPEARED BEFORE THE COURT TO file for divorce, we both made the mistake of thinking that we needed to appear at the government building in Cambridge where we signed our marriage license three years before, the same building where the first legal same-sex marriage occurred in Massachusetts. I felt a pang of shame as we walked into the building, but this feeling was cut short when we realized we were at the wrong building. We needed to go to a courthouse a few miles away. Jordan offered to give me a ride. I hesitated. Scenarios ran through my mind of what might happen, of what he might say, but my desire to disentangle was greater than my fear. I used the adrenaline in my bloodstream from fear to spur me forward. I jumped in the car.

We barely spoke to each other on the way over. We barely spoke to each other when the court clerk told us that we were the last people to show up for the walk-in divorce sessions. We barely spoke to each other when the clerk gave us permission to go ahead anyway. Maybe she saw the despair on my face. We sat down in a cold, gray courtroom. The only light that illuminated the room came from a few windows that were barely able to light the space. We sat on hard wooden benches that reminded me of church pews. Clustered throughout the benches were other couples petitioning for a divorce. I saw a middle-aged woman sitting next to a stern man with a mustache. She was sobbing uncontrollably into tissues. I saw a younger woman and man across the aisle. They were dressed in designer coats. He wore a high-end silver watch and an immaculate tie. She came with a gold necklace and pearl earrings. She sat to the right of an older female attorney, and he sat to the left of an older male attorney. Separated by their two attorneys in black trench coats, they fended for whatever they had created together—money, children, pride. I saw a lesbian couple sitting in the row in front of me sharing jokes to one another under their breaths with another woman sitting between them. Perhaps they were looking forward to ending what they agreed had been an irrational decision taken together on a whim. In the first row on the right, I saw a thin woman in a faded black sweater and a dark blue beanie covered in pilling. She was sitting by herself. Just a few feet from me, I saw an elderly woman wrapped in her shawl, keeping herself warm. She was also sitting by herself, perhaps finally saying no to a decision she was forced into from her twenties.

One by one, each their own channel of tangled emotions echoing through me—anger, disappointment, depression, relief, resentment—each person in the courtroom was called up to the stand to have their divorce papers reviewed by the judge. I sat on the bench waiting, bringing myself back into focus in my own body as I ran my fingernails between the ridges of the wood, curling my fingers underneath the bulbous edge of the bench, occasionally coming up against an old dry wad of chewing gum. "Here I am," I remember thinking. "Here's my body."

Jordan and I were the last to be called. As we stood there in front of the judge, he asked us whether we knew what was in our separation agreement. He asked us whether we were in accordance with its contents. He asked if we understood that it would be irrevocable. We each replied to his questions with a stale, "Yes." I couldn't help but feel as though I had found myself in a reverse wedding ritual, undoing our vows.

Later, we walked down the hallway of the courthouse, keeping my silence, keeping my steps brisk. We awkwardly exchanged a trite platitude as I walked past the metal detector at the entrance, back into the cold, and onto the first subway train with open doors. I didn't care where it took me. I sat in silence, feeling numb, trying to focus on not letting everything inside overwhelm me.

When I finally made it to the train station near the hospital, I walked up the steps. My phone squeaked a single buzz. I pulled the phone out of my pocket. It was a final text message from Jordan. I immediately turned and dipped into the closest corner, a small empty corridor in the soot-covered subway station. I crouched down, buried my head in my knees, and wept. I wept for all the things I could've said. I wept for all the things I could've done. I wept in gratitude for not saying them. I wept in gratitude for getting this far. I wept in pain and in gratitude until the tears stopped. I let in a deep breath, unable to carry out a complex thought to its conclusion, and texted my three former cochief residents: Ayush, Nancy, and Kate. We had gotten to know each other well as we worked together so closely in our last year of neurology residency. They, too, were my tribe. Kate happened to be at the hospital. I walked over. She was able to step out from what she was doing. We had tea. The warmth of the tea was soothing. I felt the knots in my stomach begin to unravel. I thought of Rosie and her tribe. I made a point to make sure that I had regular contact with friends, with people who were within my tribe at varying distances, even if only to talk about the most ordinary of things: my morning, my commute, what I had for breakfast, the inconsistent weather. The time I invested in the presence of others was just as vital as if it were life-sustaining medication. Because it was.

In the coming weeks, I thought often of Teri and her new normal, how important it was for her to take her new reality one step at a time. I brought my attention back to the basics and to the task-conquering reflexes I had learned as an intern. I wrote out daily check boxes for the things I needed, practical things, non-negotiable things, like water and food and sleep. Just as Teri did during her marathon, I inventoried my body, paying close attention to it, as I pushed it beyond its limits. Whenever I felt distress or discomfort, and periodically for the sake of checking in and prevention, I performed body scans from head to toe to see and feel what was going on: tension in my face or shoulders; carrying my bag in a way that led to muscle cramping; or, figuring out if I was seated, standing, or sleeping in an odd posture. I also took to the gym with religious devotion, to put my body to work, to take all the energy of stress, the need to fight or flee, and put it into a space where I could give it purpose. My first day back in a full gym I felt clunky and deconditioned with pain in my joints and muscles from long periods of sitting. With each workout, I repeated the body scans, focusing on whatever muscle group I was contracting or relaxing, noticing which muscles I was using to overcompensate, feeling for when I pulled at tendons and ligaments if I were overambitious with weight and needed to scale back. Using my body in such a way was medication, too. Coming back regularly, almost every day, it became a habit, a protected time to focus on how my body felt. Though I meditated five to fifteen minutes each day, I found that what my body needed most was physical movement. Each movement pulled me out of my head and back into my own body where I felt in line with my muscles, my bones, every fiber of my being.

While caring for my body, my mind would usually trail off into recounting stories about the divorce, stories of blame, stories of weakness, tragedy, self-pity—postmortem after postmortem. At their most intense, when even writing an email became a challenge, I needed to find a way to retell these personal stories in a way that was almost automatic, requiring little effort on my part. I remembered Rosie and how she enlisted the help of Billy Joel, Bob Marley, and Bob Dylan. I created an epic playlist of songs that told my story for me, stories closest to my own, that had rhythms and instruments that were the voice that I was seeking to redefine and speak in. There are days upon days of songs about love and romance and heartbreak. I was tripping over songs that worked well, with lyrics, vocals, and instruments that matched moods and feelings that only music could convey—from Beyoncé to Bach and back again. I went on runs just to listen to entire playlists, each chord change from minor to major lifting me up and moving me one more step forward. Sia's voice in "Bird Set Free" filled me with bold streaks of lilac and black, shooting upwards across my temples at the chorus, perfumed with the smell of grape cupcake frosting and cigarette smoke. The strings of Vivaldi's *Summer* (Presto) came down like wild scribbles of amber rain, landing on my skin with the touch and colors of golden fireworks in arrays of chrysanthemum, kamuro, spider, willow, and dahlia. These were my songs of myself—for me, chosen by me.

I cultivated my own new perspective, just as drastic and unexpected a change in perspective as Rainier had experienced. A few months after the divorce, I sat with a patient who had ongoing concerns about remembering names and crippling anxiety with panic attacks. When I asked about traumatic events in his life, he told me about divorce. Caught off guard, my emotional pain emerged quickly, scrambling to break out. Time opened up and froze, and in that moment I kindly offered an opportunity for myself to decide whether to engage it head-on. I tucked the eager emotion back and wrote down a check box to mentally return to this point again later and attend to it then. I gave myself the space to experience the pain and grief without indulging in it, so I would not unintentionally develop a habit of letting these emotions run loose on their own. This was my attempt to let the heat of my limbic system settle, without aggressively pushing it away, and then later examining it with the charity of curiosity and the patient reward of insight. It was a deliberate act of closing the distance on myself after the sudden change in context and circumstances. Over time, I regained my equilibrium with greater ease, sorting through my thoughts and feelings much like I had learned to parse through my synesthetic associations. If not for the appreciation of my synesthesia's hard-earned lessons and the many people I'd reflected through me, I would not have so readily prioritized caring for myself during my divorce. I attended to myself in order to make the time to listen to my story. I had to work on getting my reality back in order. Attending to my life in the present was necessary because my life henceforth would forever be affected by how I responded to tragedy. In a way, I found the capacity to surrender and admit myself into a personal intensive self-care unit where I was my own responding clinician. I was responsible for creating and checking the boxes that helped me accomplish the essentials I needed for life, to nourish my mind and my emotions properly, to recover and rehabilitate the conglomeration of human dimensions that we call a soul, coming together haphazardly and elegantly right below the surface of our physical limitations.

Finally remaining true to myself, I delivered myself, anew, into a new world.

Central to this experience was pain. Pain has always been part of my training—not just enduring pain but growing through it while acknowledging and attending to the pain of others. Each breath, no matter how effortful or painful, served as an aid in helping me process a deeply tied and associative world that swings between subjective and objective experiences. Each breath created an opportunity to refocus, to use the voluntary control of my body to literally stimulate my parasympathetic nervous system enough to settle the sympathetic nervous system's response, to give me time—whether one more moment or a large expanse—to let my brain catch up and understand what hurt, where the pain was coming from, and figure whether the threats I was creating were in my mind or existed in the world outside of me. First, though, I had to reflect on what was going on in me, then—and only then—attend to what was reflected in me from other people. Starting with myself allowed me to think and feel, then act in kind to another.

Through their reflections and my own, I was able at last to step forward more naturally with an ardent, interconnected sincerity. Until I understood myself and my experiences within the world at large, I couldn't exist in the common space between my *self* and other *selves,* the "little self" and, rippling outward, the "big self."

Listening, observing, feeling the experiences of others—of Fiona, Rainier, Rosie, Teri, and countless others—all allowed me to open a perspective into myself so great that I was finally able to recognize myself, my own body, my own story. Each body contains its own story, its own quiet epic. Setting it free into the world requires patience and fortitude. But this is ultimately how we declare ourselves, in our own unwavering voice, ready for the world—and, with the assistance of empathy, the only way we remain open to the world.

"The question of what it means to be human," Krista Tippett

reminds us, "has now become inextricable from the question of who we are to each other." Despite the tension between estrangement and familiarity, we cannot deny that we are each other's family. You are a stranger to another person, just as much as they are a stranger to you. And yet we are programmed cognitively as well as emotionally to recognize ourselves in others. This is the heart of empathy, the secret to bridging the gap between where you or I end and where someone else begins.

Empathy is a skill that we desperately need to train as part of our emotional education. The mirror neuron system is a network that we possess and we can experience at different levels of intensity. Practice and experience help us develop heightened activity in our mirror network. Experimental evidence supports that an increase in activity occurs in the mirror network when we watch movements we have been trained in compared to movement we have not been trained in: a ballerina has heightened activity in the mirror network when she sees a plié or a pirouette. A capoeira martial artist has heightened activity when watching an aerial *piao sem mao*, a swimmer for swimming, and so on. At a neurobiological and societal level, a human can practice empathy in order to more deeply connect with another human.

When practiced in earnest, like any other endeavor, empathy can be uncomfortable, especially when we realize that we may not like the others' differences, that we may be repulsed by occupying their same position. But this is precisely the place where we will learn the most, where the treasure lies. We can overcome the discomfort by tapping into our desire to learn, to feel, to surrender to our most innocent curiosity and ask, "What matters to you? What could be one goal that we share? What resources might we share to accomplish that goal?" This can only occur and, by extension, be experienced through each of us—for us and for each other.

Empathy is not just putting yourself in the other person's

position. It's also caring enough to consider why. If we can genuinely appreciate what is motivating others to do what they are doing, we can remain open long enough to piece together their story and understand for ourselves in our own language what it is they expect to happen next, their hope for the future they envision most clearly.

To be able to wield a constant, automatic, heightened empathy, I needed to learn how to wield a heightened form of resilience through grit, patience, and empathy for myself. Compared to most people, I may encounter a greater challenge in working to cover this distance because of my mirror-touch synesthesia. But a vast learned growth in perspective helped me better negotiate my mirror-touch experience and in the process helped me figure out how I wanted to share my story.

Answers (or at the very least, clues) about other people arise from the telling of our own story. Through curiosity, creativity, and courage we can redefine and rearticulate our personal narrative. Before being more open about my synesthesia, I was driven to confirm what the trait actually was, to understand it scientifically and explore it emotionally. Through this sensory experience, I came to grasp a little more about the human condition and, because of my efforts, found hope—reassurance in the knowledge that it is possible to safely navigate the distance between ourselves and everyone else.

Exchanging our personal stories with other people allows us to enter into a kind of communion with one another, and that might mean beginning our story from our childhood, from our family history, or from however we came to arrive here. As Paulo Coelho put it, "Even if my neighbor doesn't understand my religion or understand my politics, he can understand my story. If he can understand my story, then he's never too far from me." Though we may feel singular in our individual consciousness, we are far from confined to this singular experience. We are porous, open, and free in our interconnections, and as Walt Whitman continues to remind us, we contain multitudes.

The capacity and responsibility to compose my life lay in each liberating breath. And, just like every remarkable story, my ongoing life story features conflicts, intermittent celebrations, tragedies, and moments of unexpected beauty. No matter how tragic our circumstances may be, we can rise physically or emotionally in how we choose to redefine our personal narratives, regardless of when or how our stories end. When tragedy inevitably befalls us, we are always able to reflect, to notice with clear eyes what is happening around us and within us. This ability to reflect affords us the opportunity to figure out what to do next, one hesitant step after the other, even if our feet take us in the direction of surrender. Understanding and redefining ourselves through our stories helps us take that next step toward the other and as the recipient—listening, observing, feeling, experiencing—being open to the joy and discomfort of surrendering to new perspectives and realities.

The visible and invisible connections that exist between us—be they pheromones, herd immunity, mass hysteria, yawning, social networks, the Internet—will continue to surprise us in their dominion over us to create conflict and cooperation. We can experience the full presence and magnitude of these connections between us to move from suffering to empathy to compassion to kindness—to hope. We can remain individual while simultaneously being in communion with other people.

In one of his last essays, Oliver Sacks wrote, "There will be no one like us when we are gone, but then there is no one like anyone else, ever. When people die, they cannot be replaced. They leave holes that cannot be filled, for it is the fate—the genetic and neural fate—of every human being to be a unique individual, to find his own path, to live his own life, to die his own death."

No matter our differences, no matter how we begin, we all end. In this, we are exactly the same. In our last seconds before death, lips pressed gently against the silence, we have one last moment to appreciate the divinity of the world we have been so miraculously born into. Until my own next end, I remain Joel. I am you. And we are us.

This is our story.

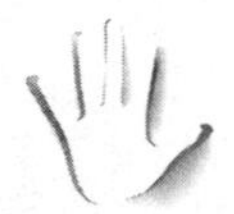

Acknowledgments

In addition to all the patients who taught and inspired me, I would like to thank my dear friends, colleagues, and others who helped support and mentor me throughout my training, from sharing a coffee to sharing a smile. I would also like to thank those who encouraged me to tell my story, including Erika Hayasaki, as well as a special thank-you to my editor, Miles Doyle, who took a chance on me and extended the initial invitation to put my experiences into words.

About the Author

Dr. Joel ('joh-EHL') Salinas is a Harvard-trained neurologist who was born with *mirror-touch synesthesia,* an extraordinary neurological phenomenon that allows him to literally feel others' emotions and physical sensations. He works at Massachusetts General Hospital as a clinician and researcher.